BORGES AND MEMORY

BORGES AND MEMORY

Encounters with the Human Brain

RODRIGO QUIAN QUIROGA

TRANSLATED BY JUAN PABLO FERNÁNDEZ
FOREWORD BY MARÍA KODAMA

The MIT Press
Cambridge, Massachusetts
London, England

This work was originally published in Spanish by Editorial Sudamericana, Buenos Aires, 2011.

Every effort has been made to contact those who hold the rights to images. Any rights holders not credited should contact the publisher.

MIT Press books may be purchased at special quantity discounts for business or sales promotional use. For information, please email special_sales@ mitpress.mit.edu or write to Special Sales Department, The MIT Press, 55 Hayward Street, Cambridge, MA 02142.

This book was set in Scala and Scala Sans by Toppan Best-set Premedia Limited. Printed and bound in the United States of America.

Library of Congress Cataloging-in-Publication Data

Quian Quiroga, Rodrigo.
[Borges y la memoria. English]
Borges and memory : encounters with the human brain / Rodrigo Quian Quiroga ; translated by Juan Pablo Fernández ; foreword by María Kodama.
    p.   cm.
Includes bibliographical references and index.
ISBN 978-0-262-01821-0 (hardcover : alk. paper)
1. Memory—Physiological aspects—Popular works. 2. Brain—Research—Popular works. 3. Memory disorders—Case studies. 4. Memory in literature. 5. Borges, Jorge Luis, 1899–1986—Criticism and interpretation. I. Title.
QP406.Q53   2012
612.8'233—dc23
2012008689

10   9   8   7   6   5   4   3   2   1

CONTENTS

Contents

FOREWORD

María Kodama

I was not surprised when I got Rodrigo Quian Quiroga's phone call asking me if he could hunt through Borges's library for any notes there might be in books related to science. Years before, scientists Roberto Perazzo and Sara Slapak had approached the Fundación Internacional Jorge Luis Borges to propose that we organize a conference on Borges and the hard sciences. The conference took place at the School of Law of the University of Buenos Aires, and the university press (EUDEBA) published the proceedings. The scientists were very interested in Borges's work in relation to the fourth dimension and to hypertext and its relation to the Internet.

Quian Quiroga explained that his specialty is neuroscience and that the story "Funes the Memorious" was closely related to his field of research. We talked about memory, a topic that fascinates me too, and one whose presence in Borges's oeuvre I had already studied—from a literary point of view, of course.

Our encounter resulted in a symposium at the Faculty of Exact Sciences of the University of Buenos Aires and in the foreword to this book, which I am delighted to write because from my earliest childhood I have been fascinated by the brain, by memory.

Borges was not a scientist, a mathematician or a physicist, but his education did include a strong grounding in philosophy, encouraged by his father, and in literature, based especially on the books by English authors that he received from his grandmother. Among his readings were works by H. G. Wells and Jules Verne, whose imagination—as powerful as Borges's—made them foresee the scientific and technical discoveries that turned dreams into reality during the twentieth century and continue to do so in the twenty-first.

According to experts, Borges's "The Garden of Forking Paths," a story from the forties, prefigured hypertext and the Internet.

Quian Quiroga's book reveals his knowledge of Borges's work and gradually, subtly, explains how it relates to neuroscience and how the one pollinates or foretells the other.

Perhaps because his work is so similar to Borges's, Quian Quiroga fathoms two of the fundamental themes running through "Funes the Memorious" that are essential for the development of humanity: abstraction and forgetting. In his *Naturalis historia*, Pliny the Elder had already referred to people with prodigious memory; what to Pliny is a wondrous gift, to Borges, who thought deeper about the subject, may turn into a terrible curse.

To Quian Quiroga, the cyberworld in which most people are immersed nowadays is sometimes crammed with undigested information, just like Funes's brain. To Quian Quiroga our world occasionally serves us with a hodgepodge of ideas, images, bits of news, coming at us relentlessly, incoherently, and we end up in a virtual world that alienates us ever more and takes us ever further away from what really makes us human: reflection, the ability to distance ourselves from our surroundings so we can calmly ponder and try to understand even a tiny point, the universe.

Most people, I believe, would find it difficult to guess what a scientist does day to day. The first image that comes to mind is of an untidy, chaotic person, always lost in thought, absentminded; someone alien to the surrounding, mundane reality, who does not realize if it is raining, if it is Tuesday, if it is a national holiday, or if his bus has just passed by; someone who spends whole days filling blackboards with theories and formulas in the search of a "Eureka," a discovery that will add a bit, however tiny, to our knowledge. But this expression of Archimedes is very rare in the life of a scientist.[1] In fact, in most cases, even after years and years of

1. The circumstances that led Archimedes to cry "Eureka" are a tasty bit of science lore. Legend has it that king Hiero of Syracuse commissioned a wreath from a goldsmith and upon receiving it became suspicious that

research, such a moment never arrives. Isaac Asimov, the extraordinary biochemist and science fiction writer, once said that the expression that accompanies a discovery is usually not "Eureka" but "This is funny . . .". In other words, this moment of ecstasy that should get us running naked through the streets of Syracuse may end up being just a moment of doubt, an initial enigma that will be resolved only after years of research.

What is it, then, that makes scientists wander about in a universe of ideas and experimentation? It may be the search for knowledge or, in more mundane terms, simple curiosity. Nagging questions; the pressing need to figure something out and the inability to do anything else until the answer is found; the tingling feeling that a discovery may be just around the corner; the intuition that a puzzle

he may have been swindled; he asked Archimedes to determine if the wreath was made of pure gold or if cheaper metals had been mixed in during manufacture. Of course Archimedes could not melt the wreath to measure its volume (and thus estimate its density and compare it to that of pure gold). But one day, while taking a bath, he noted that the water level in the tub rose as he dipped himself, and from this observation he deduced that the ornately shaped wreath must have the same volume as the liquid it would displace if submerged. According to the legend, as soon as he realized this he set off naked through the streets of Syracuse shouting "Eureka" ("I found it!" in ancient Greek). As with every other legend, there are several people (including Galileo) who doubt its veracity and believe it more likely that Archimedes said "Eureka" when he discovered the relation between the upward lift force experienced by a submerged body and the volume of liquid it displaces. This latter result is no doubt more relevant scientifically, but it lacks the magic of the original.

is starting to take shape, until eventually one reaches the answer and feels the thrilling joy of understanding.

One can then ask whether scientists, embarked upon their personal quests—their quixotic endeavors—spend their time just thinking. Not really. The life of a scientist is generally more humdrum and may involve repeating an experiment for the nth time to check the validity of a result, or analyzing data in a computer to extract some additional information. A sociologist may spend a lot of time planning surveys and analyzing statistics, a biologist preparing samples and dealing with pipettes, a mathematician varying systematically the parameters of a model, and a neuroscientist recording the activity of hundreds of neurons and crunching terabytes of data. This may sound somewhat boring, but if there is a worthwhile question lurking behind it, the routine becomes fascinating, and from those quotidian tasks the scientist weaves an elaborate plot to get ever closer to the answer of the problem that has resulted in so much lost sleep.

In my own particular case, this plot has to do with the functioning of the brain (though not the *whole* brain, since it is impossible for a single person to encompass the knowledge gathered in even a single branch of science). And in my quest to understand different aspects of how the brain works—and more specifically of how memory, the topic of this book, works—it is rare, very rare, to come by a "Eureka." Problems are usually left open, answers usually lead to further questions, and the final solution is almost always elusive. But perhaps our obstinate perseverance may be nothing more than the knowledge that, at least subconsciously, the pleasure is not in finding the answer but in searching for it. And without blushing I dare say that my search, shared with many colleagues, may well be the most interesting of all. Thus, beyond the fact that the human brain is the most complex and elusive mystery of science, the truth is that the quest to understand the brain is ultimately the quest to understand ourselves. And although we know fairly little, most of what we do know has been discovered in the last few decades. This is the ideal time to study the brain, just as the era of Galileo and Newton was ideal to study the motion of bodies and Maxwell's to study electricity and magnetism.

Nowadays we have at our disposal sophisticated equipment and advanced methods to analyze massive amounts of complex data. We also have access to information that we could not have dreamed of just a couple of decades ago. What was science fiction a few years back is becoming fact at a vertiginous pace. However, in our mad dash to understand ever more about the behavior of the brain we tend to forget that this search is not exclusively ours, of researchers with sophisticated labs, but has also been undertaken by many great thinkers: from the ancient Greek philosophers to the Cartesian rationalists, the British empiricists, and the nineteenth-century pioneers of modern psychology, along with other brilliant intellectuals who defy any categorization, like Jorge Luis Borges, who reached astounding conclusions guided only by his reasoning and his prodigious imagination.

It is not uncommon for a scientist to be interested in Borges, especially if (like me) he had the good fortune to study at the Faculty of Exact Sciences of the University of Buenos Aires (UBA). Deep and varied connections sprang out as we read about the aleph—the cardinality of infinity, studied in advanced calculus—about forking paths that lead to parallel universes—as in some interpretations of quantum mechanics—or about an infinite library that in the end turns out to have the same contents as a single "book of sand," whose number of pages is a continuum.

Like many others, I discovered Borges as a teenager and was fascinated by the mathematical precision with which he describes what defies every logic, with the way he starts from seemingly irrefutable premises—often reinforced by obscure or even blatantly apocryphal quotations—to lead us inexorably into unreal worlds as though we were hallucinating or dreaming, living in a *fantastic realism* where everything is possible and ideas rule above all else. Many years later I rediscovered a story of his, "Funes the Memorious," that had the perfect words to express the results of my research and which with astonishing clarity ended up sorting the pieces of the puzzle I had been working on. In brief, together with colleagues at Caltech and UCLA I was lucky enough to find neurons in the human brain that respond to abstract concepts, ignoring particular details. These neurons play a key role in turning what we perceive—what we see, touch, hear—into long-term memories that we will recall years later. When we generate memories we seek to abstract, to synthesize concepts. Usually we would rather not memorize details, lest we end up like Funes. And, just as in one of Borges's stories, there I was, the scientist, obsessed with trying to understand discoveries whose interpretation was already there, in a book written more than half a century before, a book that I had read as a youngster and that lay lost in my memory.

Carried away by the game of conceiving the plot of this apocryphal story—which, after all, is the story of this book—I imagine a Borgesian universe in which the main character is a monk in a monastery library—as in Umberto Eco's *The Name of the Rose*—reading a book that was believed lost; or a Persian vizier out of *The Thousand*

and One Nights, finding his truth in a story told by a traveler; or perhaps a *compadrito* from the south of Buenos Aires in the early twentieth century who ends up challenging to a duel the songster who recites, with as much skill as little mercy, the verses he has long been searching for. In all these plots there is, I would say, only one way to continue the story: the scientist, whatever form he takes, strives to understand how the author of the lost book, or the traveler, or the strummer, chanced upon the answer to his question.

My search for further information about the readings or the facts that triggered in Borges the idea of Funes, his interest in memory and the workings of the brain, led me to contact his widow, María Kodama. Science is a discipline in which good luck often turns out to be as important as intelligence, creativity, and persistence. I had the good fortune that María took an interest in this parallel between Borges and memory, shared countless stories with me, and granted me access several times to the books in Borges's personal library. Just as in "The Circular Ruins" the man who creates another man in his dreams realizes that he himself is being dreamed by somebody else, I found, to my amazement, that, in the same way that Borges had perhaps already dreamed results like the ones I was lucky enough to discover in my research, other men before him—William James, Gustav Spiller, or John Stuart Mill—had perhaps also dreamed a story like Funes's. I have no desire to question Borges's originality (I am far from being the *compadrito* seeking a duel); rather I am an astonished scientist who wants to know better the man who helped him organize his ideas. Beyond the fact that the ideas behind "Funes the Memorious" had been circulating since the late nineteenth century or even before, no one can doubt Borges's genius in buffing them to perfection in a marvelous story.

Borges was not a scientist, but his passion for literature and philosophy led him to study psychology and the workings of the mind—and here I use "mind" instead of brain to highlight a more philosophical connotation. I took the opposite road: starting from current open questions in neuroscience I was led by Borges's readings to the foundations of psychology and philosophy. To peruse the books in Borges's library, as I did thanks to María Kodama's kindness, was like having an intimate conversation with him, from which I got an idea of his interests and started to grasp his thoughts. Almost without realizing it, I ended up exploring a terrain so ancient that it turned out to be novel. In today's scientific world, in which everything occurs at dizzying speed, where we barely manage to digest the information that reaches us, these encounters with Borges gave me a much-needed chance to take a pause to think in depth and debate (in my mind) with Descartes, Bishop Berkeley, and James. How misguided we scientists are when we think we are the first to deal with the big questions! We are simply sharpening and rephrasing the same questions that Aristotle asked himself more than two millennia ago.

This (anachronistic) search for Funes's roots and his relation to the principles of neuroscience is in fact the topic of this book, which started as a brief scientific paper that went through many drafts and at the end was left too brief, barely a sketch of an idea.[2] Each pruning of the manuscript made it lose nuances to which I now hope to do justice. This is neither a book about Borges nor a

2. Rodrigo Quian Quiroga, "In Retrospect: *Funes the Memorious,*" *Nature* 463 (2010): 611.

textbook on memory; rather, it stems from my urge to tell a story that I find fascinating. The urge is such that I cannot do anything else until I finish; so fascinating that the book almost writes itself, as though I were narrating it to a friend with no knowledge of scientific jargon but who shares my curiosity and interest about Borges and the workings of the brain. I am not trying to force a link or suggest that Borges foresaw modern neuroscience. Neither am I attempting to overpraise Borges or judge him beyond his perfect prose and his extraordinary intuition in dealing with a topic as engrossing as memory. Borges is perhaps the catalyst that persuaded me to tell a story, a story that must inevitably begin with Funes the memorious . . .

Kleve, August 2010

FUNES AND OTHER CASES OF EXTRAORDINARY MEMORY

June 7, 1942, was a Sunday like any other amid the altered routine of the Second World War. The front page of the newspaper *La Nación*[1] reported on the British onslaught, which continued with a bombing campaign over the German industrial area in the Ruhr. On the same page one could read about the casualties inflicted on the Japanese fleet at Midway and about British infantry tanks' attacking German positions in the desert. Pages 5 and 6 of the paper, in between advertisements for Eno's "Fruit Salt" (a digestive aid selling at $0.70 per vial) and Fernet Branca (a beverage that should be brought home as one brings a friend), give an account of an earthquake without victims in Mendoza and announce that tire factories can start restoring used tires. In sports, Argentinos

---

1. One of the main newspapers in Argentina.

Juniors beat Sportivo Alsina by 4 goals to 1 in their campaign to reach the premier league, and the entertainment pages promote *Pirates of the Caribbean*, in Technicolor, and a new movie starring Olivia de Havilland and Henry Fonda at $1.50 a *superpullman* seat. June 7, 1942, a day like any other according to *La Nación*, except for a short story appearing in the Arts and Letters section that would turn this issue of the newspaper into a historic document. The first page of this Sunday supplement features a story by Stefan Zweig; the second page contains an essay by Ernesto Sabato praising Galileo; and on the third page, almost hidden in plain sight, for the first time appears "Funes the Memorious," Jorge Luis Borges's monumental short story, with an illustration by Alejandro Sirio.

"Funes the Memorious" tells the vicissitudes of Ireneo Funes, a peasant from Fray Bentos, who after falling off a horse and hitting his head hard recovers consciousness with the incredible skill—or perhaps curse—of remembering absolutely everything.

Says Borges of Funes:

Nosotros, de un vistazo, percibimos tres copas en una mesa; Funes, todos los vástagos y racimos y frutos que comprende una parra. Sabía las formas de las nubes australes del amanecer del 30 de abril de 1882 y podía compararlas en el recuerdo con las vetas de un libro en pasta española que sólo había mirado una vez y con las líneas de la espuma que un remo levantó en el Río Negro la víspera de la acción del Quebracho.[2]

---

2. Jorge Luis Borges, "Funes el memorioso," in *Obras completas* (Buenos Aires: Emecé, 2007), vol. 1, pp. 583–590; subsequent quotations from this story are from the same source. Except where otherwise specified, all English translations of Borges in this book are by Juan Pablo Fernández.

FIGURE 1.1

Page 3 of the Arts and Letters section of *La Nación* of June 7, 1942, where "Funes the Memorious" was first published.

[We, at a stroke, perceive three cups lying on a table; Funes would see all the shoots and clusters and fruit comprised by a vine. He knew the shapes of the southern clouds at dawn on April 30, 1882, and could compare them in his memory with the streaks on a book of Spanish cover that he had seen only once and with the swirls on the foam raised by an oar in the Río Negro on the eve of the battle of the Quebracho.]

Jorge Luis Borges (1899–1986) has received universal acclaim for the depth with which he approached matters of philosophic and scientific import in his writings. In Borges's hands, the topic of infinity comes alive either as a point that contains the universe ("The Aleph"), impregnable labyrinths ("The Two Kings and the Two Labyrinths"), a library that is eternally repeated ("The Library of Babel"), stories that subdivide into innumerable possibilities ("The Garden of Forking Paths"), or an imperial map so perfectly detailed that it ends up having the size of the empire itself ("Of Rigor in Science"). In "Funes the Memorious," a story of barely 12 pages that was eventually published as part of *Ficciones* (1944), Borges again plays with the infinite in a context no less fascinating: the vast labyrinths of memory and the consequences of having an unlimited capacity to remember.

Funes is first mentioned in an obituary of James Joyce, "A Fragment on Joyce," published in 1941 in the magazine *Sur*.[3] There,

3. I am not the first to reflect on the roots of "Funes the Memorious" and its possible interpretations. In fact, I would like to give credit to those who have previously written on this topic (many of them cited in subsequent footnotes), and although any list I make will end up being unfair, since I will surely forget more than one reference, I would like to mention the essays by Víctor Zonana ("Memoria del mundo clásico en 'Funes el

with some measure of sarcasm, Borges says that to read straight through a "monster" like Joyce's *Ulysses*—a 400,000-word reconstruction of a single day in Dublin—requires another monster able to remember an infinite number of details. The strange thing about the obituary is that Borges barely refers to Joyce or his work and instead describes Ireneo Funes, the main character of the story he was writing at the time.

Entre las obras que no he escrito ni escribiré (pero que de alguna manera me justifican, siquiera misteriosa y rudimental) hay un relato de unas ocho o diez páginas cuyo profuso borrador se titula "Funes el memorioso". . . . Del compadrito mágico de mi cuento cabe afirmar que es un precursor de los superhombres, un Zaratustra suburbano y parcial; lo indiscutible es que es un monstruo. Lo he recordado porque la consecutiva y recta

memorioso'" [Remembrance of the classical world in "Funes the Memorious"], whose introduction includes an excellent summary of related work); Roxana Kreimer ("Nietzsche, autor de 'Funes el memorioso': Crítica al saber residual de la modernidad" [Nietzsche, author of "Funes the Memorious": A critique of modernity's residual knowledge]); Eduardo Mizraji ("Memoria y pensamiento" [Memory and thought], among other essays in the book *Borges y la ciencia* [Borges and science]); Patricia Novillo-Corvalán ("James Joyce, Author of 'Funes the Memorious'"); Carlos Baratti ("'Funes el memorioso': Ficción que invita a reflexionar acerca de la neurobiología de la memoria" ["Funes the Memorious": A fiction that invites reflection on the neurobiology of memory]); and the books by Iván Izquierdo (*El arte de olvidar* [The art of forgetting]), Guillermo Martínez (*Borges y la matemática* [Borges and mathematics]), and Diego Golombek (*Cavernas y palacios: En busca de la conciencia en el cerebro* [Caverns and palaces: Searching for consciousness in the brain]). Funes is, I would say, a classic reference in any book by an Argentine author on the topic of memory.

lectura de las cuatrocientas mil palabras de *Ulises* exigiría monstruos análogos.[4]

[Among the works that I have not written and will never write (but that somehow justify me, in however mysterious and rudimentary a way) there is a short story, some eight to ten pages long, whose copious draft is entitled "Funes the Memorious." . . . Of the magical *compadrito* of my story I can state that he is a precursor to supermen, a suburban, incomplete Zarathustra; what cannot be denied is that he is a monster. I have remembered him because a straight, uninterrupted reading of *Ulysses*'s four hundred thousand words would require similar monsters.]

In the preface to "Artifices," the second part of *Ficciones*, Borges argues that "Funes the Memorious" is a long metaphor of insomnia. In fact, toward the end of the story he mentions that Funes found sleeping difficult, because to sleep is to get distracted from the world. Borges gives more details on the way he conceived Funes during his own sleepless nights (perhaps during a sticky summer night at the *quinta* in Adrogué), in an interview published in the United States:

When I suffered from insomnia I tried to forget myself, to forget my body, the position of my body, the bed, the furniture, the three gardens of the hotel, the eucalyptus tree, the books on the shelf, all the streets of the village, the station, the farmhouses. And since I couldn't forget, I kept on being conscious and couldn't fall asleep. Then I said to myself, let us suppose there was a person who couldn't forget anything he had perceived, and it's well known that this happened to James Joyce, who in the course of a single day could have brought out *Ulysses*, a day in which thousands of

4. Jorge Luis Borges, "Fragmento sobre Joyce," in *Jorge Luis Borges en Sur, 1931–1980* (Buenos Aires: Emecé, 1999), pp. 167–169.

things happened. I thought of someone who couldn't forget those events and who in the end dies swept away by his infinite memory. In a word that fragmentary hoodlum is me, or is an image I stole for literary purposes but which corresponds to my own insomnia.[5]

Already in the literature of the first millennium there are references to people with prodigious memory, particularly in the *Naturalis historia* (*Natural History*) of Pliny the Elder (Gaius Plinius Secundus, 23–79 A.D.), a sort of encyclopedia that in 37 books describes everything from the geography, science, and technology to the agriculture, medicinal herbs, and insects of ancient Rome. In chapter 24 of book VII, on the topic of memory, Pliny mentions king Cyrus of Persia, who knew the names of all his soldiers; Scipio, who knew the names of all in Rome; Cineas, king Pyrrhus's ambassador, who learned the names of all the Roman senators just one day after arriving in Rome; Mithridates Eupator, who administered justice in the 22 languages spoken in his empire; Simonides, inventor of mnemonics; or Charmadas the Greek, who could recite by heart any book from a library as though he were reading it.[6]

Pliny considers it a blessing to possess an extraordinary memory. In fact, he starts chapter 24 of book VII saying: "Memoria

5. *Jorge Luis Borges: Conversations*, ed. Richard Burgin (Jackson: University Press of Mississippi, 1998), p. 166. This passage of the interview has been cited by Patricia Novillo-Corvalán in "James Joyce, Author of 'Funes the Memorious'."

6. Most of these characters had been earlier described by Cicero in his *Tusculan Disputations*. See Cicero, *Tusculan Disputations*, rev. ed., trans. J. E. King (Cambridge, MA: Harvard University Press; London: Heinemann, 1960).

FIGURE 1.2

Title page of the first volume of a 1669 edition of Pliny's *Naturalis historia*.

necessarium maxime vitae bonum cui praecipua fuerit, haut facile dictu est, tam multis eius gloriam adeptis [As to memory, the boon most necessary for life, it is not easy to say who most excelled in it, so many men having gained renown for it]."[7]

Pliny also describes the fragility of memory, arguing that it can be lost, in whole or in part, due to illness, injury, and even panic.

7. Pliny, *Natural History*, vol. 2, trans. Harris Rackham (Cambridge, MA: Harvard University Press, 1942; London: Heinemann, 1947), pp. 562, 563.

As an example he tells the story of a man who lost the capacity to name letters after being struck by a stone, and of another who forgot certain people after falling from a roof. He also mentions Messala Corvinus, the orator, who lost recollection of even his own name.

Borges, it is known, was fascinated by encyclopedias and by the *Naturalis historia*[8] (perhaps the first encyclopedia in history), which in fact he mentions in "Funes the Memorious": Funes asks the narrator (Borges) for any Latin text, and Borges obliges with volume VII of Pliny's encyclopedia and Quicherat's *Thesaurus*, just so the rube will be rudely disappointed upon finding out that one cannot learn such a complicated language using only a book and a dictionary. On their next meeting, however, Funes welcomes Borges by reciting, mockingly, in perfect Latin: "ut nihil non iisdem verbis redderetur auditum" (which, literally translated, means: "Nothing that has been heard can be repeated with the same words").[9]

8. Upon receiving the Cervantes Prize in 1979 ("a generous blunder that I shamelessly accept," he said), Borges commented in an interview that with the prize money—a million pesetas, to be shared with Spanish poet Gerardo Diego—he planned to buy the Espasa Calpe encyclopedia, which he eventually received as a present from the publishers. Borges also had several editions of the *Naturalis historia* in his library, along with a 1907 edition of Sir Francis Galton's *Inquiries into Human Faculty and Its Development*, on whose final page Borges transcribed chapter 24 of book VII of the *Naturalis historia* in the original Latin (along with a French translation on the first page).

9. In the context of Pliny's paragraph this phrase can also be translated as "[Through memory] it is possible to repeat with the same words what has been heard."

Through Funes, Borges, just like Pliny, enters the realm of memory, though his reaction differs from the Roman's in a crucial regard: while Pliny considers it a virtue to have a prodigious capacity to remember, Borges looks beyond and argues that an extraordinary memory can become a curse. Says Funes, midway through the story:

Más recuerdos tengo yo solo que los que habrán tenido todos los hombres desde que el mundo es mundo. . . . Mi memoria, señor, es como un vaciadero de basura.

[I alone have more memories than all men may have ever had since the world exists. . . . My memory, sir, is like a rubbish heap.]

Given their historical significance, Pliny's stories are of undeniable value. It is, nonetheless, impossible to judge their veracity, and in fact the characters described in the *Naturalis historia* seem more legendary than real (perhaps arousing Borges's curiosity even more). To a large extent this is due to the fact that many of Pliny's descriptions are based on word-of-mouth information, inevitably altered in the telling. For example, when he describes cases of astonishing eyesight in chapter 21 of book VII, Pliny writes that Homer's *Iliad* was written in such small script that the complete manuscript could fit in a nutshell; he also mentions a man called Strabo, who could recognize objects 135 miles away and who, during the Punic Wars, could sight and even count the enemy ships docked in Carthage from a promontory in Sicily.

The first properly documented case of extraordinary memory is that of Solomon Shereshevskii, studied by the celebrated Russian psychologist Alexander Luria starting in the 1920s. As Luria reports in his book *The Mind of a Mnemonist: A Little Book about a Vast Memory*, subject S. (as he refers to Shereshevskii to protect his name), unlike everyone else, had to make an effort if he wished to forget something. As we shall see in the following chapters, Shereshevskii possessed a very strong synesthesia—an involuntary link between different senses, like associating numbers with colors— that gave his memories a much richer content and thus made them easier to recollect. These associations, as well as the use of simple mnemonics, allowed Shereshevskii to remember long sequences of numbers and letters many years after first hearing them. After studying Shereshevskii for more than 30 years, Luria confessed his inability to find a limit to S.'s memory, a surprising statement considering that it comes not from an amateur but from one of the foremost psychologists of his time.

FIGURE 1.3
Alexander Luria (1902–1977), Friedrich Nietzsche (1844–1900), and William James (1842–1910).

There are clear parallels between Shereshevskii and Funes, despite the fact that the former trained his memory based on his synesthesia while for the latter to remember everything was completely natural. It is, however, unlikely that Borges knew of Luria's work, since Luria published his book (in English) only in 1968, more than 25 years after Borges wrote the story of Funes.

"Funes the Memorious" shows Nietzsche's influence as well (as Roxana Kreimer describes in an interesting essay);[10] in particular, Borges calls Funes "a precursor to supermen, a suburban, incomplete Zarathustra." In a brilliant piece on the importance of forgetting, Nietzsche writes:

Imagine the most extreme example, a human being who does not possess the power to forget, who is damned to see becoming everywhere; such a human being would no longer believe in his own being, would no longer believe in himself, would see everything flow apart in turbulent particles, and would lose himself in this stream of becoming; like the true student of Heraclitus, in the end he would hardly even dare to lift a finger. All action requires forgetting, just as the existence of all organic things requires not only light, but darkness as well.[11]

10. Roxana Kreimer, "Nietzsche, autor de 'Funes el memorioso': Crítica al saber residual de la modernidad" (Nietzsche, author of "Funes the Memorious": A critique of modernity's residual knowledge), in *Jorge Luis Borges: Intervenciones sobre pensamiento y literatura* (Buenos Aires: Paidós, 2000).

11. Friedrich Nietzsche, "On the Utility and Liability of History for Life," in *Unfashionable Observations*, trans. Richard T. Gray (Stanford: Stanford University Press, 1995). In this essay Nietzsche refers to forgetting in a historical context, suggesting that man should not tie himself to the prejudices of History (a fundamental requirement for the creation of his famous "superman").

Borges's fascination with the mind (in this philosophical context I again use "mind" rather than "brain," though I make no distinction between the two) probably came from his father, a lawyer and psychology professor who introduced him to authors such as William James, considered by many to be the father of modern psychology. In *The Principles of Psychology* (1890), one of his foremost works, James says this about memory:

If we remembered everything, we should on most occasions be as ill off as if we remembered nothing. . . . "The paradoxical result [is] that one condition of remembering is that we should forget. Without totally forgetting a prodigious number of states of consciousness, and momentarily forgetting a large number, we could not remember at all."[12]

The relation to Funes, Shereshevskii, and Nietzsche is fascinating. Luria, for example, writes that Shereshevskii "was quite inept at logical organization." Borges, in turn, says that Funes

había aprendido sin esfuerzo el inglés, el francés, el portugués, el latín. Sospecho, sin embargo, que no era muy capaz de pensar.

[had effortlessly learned English, French, Portuguese, Latin. I suspect, however, that he was not very capable of thinking.]

Again: I do not refer to Joyce, Pliny, Luria, Nietzsche, and James so as to question the originality of Borges's story. On the contrary,

12. William James, *The Principles of Psychology*, authorized ed., vol. 1 (New York: Henry Holt, 1890; repr., New York: Dover, 1950), pp. 680–681. The second half of the quotation is itself a quotation: Théodule Ribot, *Les maladies de la mémoire* (Paris: Librairie Germer Ballière, 1881), p. 46.

these parallel writings provide a philosophical and scientific foundation in which Borges may have found part of his inspiration. Leaving aside the issue of whether Borges knew of Luria's studies or not—I believe not—I cannot help noticing the uncanny lucidity with which he treats a topic as complex as memory in the context of a short story.

Going back to Funes and other people with extraordinary memory, we must mention Borges himself, who could quote whole passages in Spanish, English, German, and Anglo-Saxon, among other tongues. Though it is possible that blindness may have contributed to his incredible memory (not being distracted by visual stimuli, he could focus, like Democritus before him,[13] on his thoughts and the stream of his remembrance), Borges's youthful realization that he, like his father, would lose his eyesight took him on a monumental quest for knowledge while he could still see. María Kodama remembers that, on one of her first encounters with Borges, he asked her to find an excerpt from a book. The fragment, the writer said, was on an odd-numbered page near the middle of the book. Kodama started to read a page at random and Borges, amazingly, guided her to the right page even though he had been blind for many years and—as he jotted on the first page—had read the book in 1916, decades before this encounter with Kodama.

13. Democritus is known for conceiving atomic theory; legend has it that he gouged his eyes out in his garden so that contemplation of the external world would not disturb his meditations.

THE LIBRARY OF BABEL

Borges saw himself less as an extraordinary writer than as a passionate reader. He starts his poem "A Reader" by saying:

Que otros se jacten de las páginas que han escrito;
a mí me enorgullecen las que he leído.[1]

[Let others boast of the pages they have written;
I take pride in those that I have read.]

Of course, whoever has read Borges will disagree with this assessment of his own writing, but one must acknowledge that his prolific reading had an enormous impact on his work. He grew up reading the books in his father's library, a room that during his childhood

---

1. In Jorge Luis Borges, "Elogio de la sombra," in *Obras completas* (Buenos Aires: Emecé, 2007), vol. 2, pp. 450–451.

appeared almost infinitely large and whose glass-lined shelves contained thousands of volumes.[2] It housed many encyclopedias, along with books on psychology and a great variety of Argentine texts and English literature. Borges envisioned paradise as a sort of library, perhaps not with an infinite number of hexagonal galleries and 410-page books with random inscriptions, as in the library of Babel, but certainly with a myriad of books on every conceivable topic, where he could lose himself reading *The Thousand and One Nights*, a short story by Kipling, or philosophical disquisitions by Berkeley and Hume. This is how he, as a child, viewed his father's library, and later the National Library, where he was appointed director in 1955, ironically the year in which his blindness advanced to the point that he could never read or write again. Borges refers to this circumstance in his famous "Poem of the Gifts":

Nadie rebaje a lágrima o reproche
esta declaración de la maestría
de Dios, que con magnífica ironía
me dio a la vez los libros y la noche.[3]

[Let no one debase to tears or a reproach
this statement of the mastery
of God, who with magnificent irony
at once gave me books and the night.]

2. Jorge Luis Borges, "An Autobiographical Essay," in *The Aleph and Other Stories*, ed. and trans. Norman Thomas di Giovanni in collaboration with the author (New York: E. P. Dutton, 1970).

3. In Jorge Luis Borges, *El otro, el mismo* (Buenos Aires: Emecé, 1969); *The Self, the Other*, in *Selected Poems, 1923–1967*, ed. Norman Thomas di Giovanni (New York: Delacorte Press/Seymour Lawrence, 1972).

Books were of such importance to Borges that in his poem "My Books" he writes:

Mis libros (que no saben que yo existo)
son tan parte de mí como este rostro
de sienes grises y de grises ojos
que vanamente busco en los cristales
y que recorro con la mano cóncava.
No sin alguna lógica amargura
pienso que las palabras esenciales
que me expresan están en esas hojas
que no saben quién soy, no en las que he escrito.
Mejor así. Las voces de los muertos
me dirán para siempre.[4]

[My books (who do not know that I exist)
are as much a part of me as this face
with gray temples and gray eyes
that I vainly search for in the looking-glass
and whose contour I explore with my concave hand.
Not without some understandable grief
I think that the essential words
that express me are there, on those pages
that do not know who I am, not in the ones I have written.
Better that way. The voices of the dead
will tell me forever.]

Borges used to write minuscule notes in his books, usually on the first or last page. (Later, after he became blind, he would ask

---

4. In Jorge Luis Borges, *La rosa profunda* (Buenos Aires: Emecé, 1975); *The Unending Rose*, in *The Gold of the Tigers*, trans. Alastair Reid (New York: E. P. Dutton, 1977).

whoever was reading to him to write them.) He would use them to note passages that he would like to reread, and often to relate them to passages in other books. The Fundación Internacional Jorge Luis Borges, run by María Kodama, treasures Borges's private library, his universe. The fascination produced by studying those books is perhaps like the one felt by an archaeologist who has just found a hidden chamber in a temple or an engraving from an extinct civilization. These books are a testimony of Borges's brilliant mind, his readings and the fragments therein that caught his attention or perhaps triggered the idea for a story.

FIGURE 2.1

The author with María Kodama and an issue of *Martín Fierro* at the Fundación Internacional Jorge Luis Borges. (Photograph by Roy Gorfinkel, originally published in the newspaper *Perfil*, whose kindness we acknowledge.)

On the first page of an 1898 English edition of Leibniz's *Monadology*, Borges wrote in longhand his famous "Argumentum ornithologicum," which he dated "Buenos Aires, May 23, 1951":

Cierro los ojos y veo una bandada de pájaros. La visión dura un segundo o acaso menos; no sé cuántos pájaros vi. ¿Era definido o indefinido su número? El problema involucra el de la existencia de Dios.[5]

[I close my eyes and see a flock of birds. The vision lasts for a second or maybe less; I cannot say how many birds I saw. Was their number definite, or indefinite? The problem involves that of God's existence.]

The title of the story is evidently a pun on the "Ontological Argument" proposed by St. Anselm in the eleventh century to prove the existence of God (an argument that was later strongly criticized by Thomas Aquinas). Beyond the historical value of the original, perhaps the most interesting aspect of this document is a note by Borges, "The ontological proof, 274," which refers to a page in an appendix describing proofs of the existence of God (with variations on the ontological argument by Spinoza, Descartes, and St. Anselm himself) that may have been the spark that ignited Borges's story.

Borges was not a mathematician or a scientist. He excelled in the humanities (literature, languages, philosophy) but also approached the sciences when they aroused his imagination and offered him themes and topics for his stories. The very titles of the books in his library are revealing: they range from the nature of time (Alexander

5. Jorge Luis Borges, "Argumentum Ornithologicum," in *Obras completas*, vol. 2, p. 198.

Gunn, 1929) and the fourth dimension (Hinton, 1939, among others) to the analysis of mind (Bertrand Russell, 1921).[6]

Borges's approach to science has been already described, notably in a collection of essays entitled *Borges and Science*[7] and in *Borges and Mathematics* by Argentine writer and mathematician Guillermo Martínez. Infinity is a recurring theme in Borges's work. Two essays in *Discussion*,[8] "The Perpetual Race of Achilles and the Tortoise" and "Vicissitudes of the Tortoise," are about Zeno's famous paradox: Achilles tries to catch up with a tortoise running 10 meters ahead; when Achilles runs the 10 meters, the tortoise has advanced one meter; when Achilles advances that meter, the tortoise is ahead by 10 centimeters; when Achilles covers those 10 centimeters the tortoise is one centimeter in front; and so on. It seems as though Achilles, in spite of running ten times as fast as the tortoise, is never able to surpass it. The paradox is based on the infinity of the continuum, as one can keep on forever dividing the distance run by the tortoise. A similar principle is used in "The Book of Sand" (1975), where the number of pages in a book is a continuum. In his discussion of Zeno's paradox, Borges mentions arguments

6. John Alexander Gunn, *The Problem of Time: An Historical and Critical Study* (London: George Allen and Unwin, 1929); C. Howard Hinton, *The Fourth Dimension* (London: George Allen, 1912); Bertrand Russell, *The Analysis of Mind* (London: George Allen and Unwin, 1921).

7. Sara Slapak, ed., *Borges y la ciencia* (Buenos Aires: Eudeba, 1999). The book includes an essay by Eduardo Mizraji entitled "Memoria y pensamiento" ("Memory and Thought") that, in synergy with the present book, makes a parallel between Funes, the generation of concepts, and memory.

8. Jorge Luis Borges, *Discusión* (Buenos Aires, 1932).

brought forth by John Stuart Mill, William James, and Bertrand Russell, among others, and the books by these authors in his personal library have notes by him, with references to the pages in which the paradox is discussed. In my opinion, however, Borges's most interesting treatment of infinity appears in "The Aleph" (1949), which refers to a point in Carlos Argentino Daneri's basement that contains the whole universe (*"the only place on earth where all places are—seen from every angle, each standing clear, without any confusion or blending"*). In this story, the narrator (Borges) mentions his "writer's desperation" as he tries to express in words the idea of infinity, but in the end he masterfully describes the Aleph with a one-page enumeration:

Vi el populoso mar, vi el alba y la tarde, vi las muchedumbres de América, vi una plateada telaraña en el centro de una negra pirámide, vi un laberinto roto (era Londres) . . . sentí vértigo y lloré, porque mis ojos habían visto ese objeto secreto y conjetural, cuyo nombre usurpan los hombres, pero que ningún hombre ha mirado: el inconcebible universo.[9]

[I saw the teeming sea, I saw dawn and twilight, I saw the multitudes of the Americas, I saw a silvery cobweb at the center of a black pyramid, I saw a broken labyrinth (it was London) . . . I felt vertigo and I wept, because my eyes had seen that secret and conjectural object whose name is arrogated by men, but which no man has ever seen: the inconceivable universe.]

9. Borges himself translated the story with di Giovanni. It appears in Jorge Luis Borges, *The Aleph and Other Stories: Together with Commentaries and an Autobiographical Essay*, ed. and trans. Norman Thomas di Giovanni in collaboration with the author (New York: E. P. Dutton, 1970), pp. 15–32.

His choice of the first letter (aleph, $\aleph$) of the Hebrew alphabet to refer to infinity is not casual, since in mathematics that symbol represents "the cardinality of transfinite numbers," or, in other words, the size of an infinity. One intuitively feels that there are more rational numbers—those that can be expressed as quotients of integers, like 18.125 or 477.8—than integers, and more integers than prime numbers—those that cannot be written as products of other numbers. However, as mathematician Georg Cantor proved in the late nineteenth century, those three sets have exactly the same number of elements, the same cardinality, which is called aleph-null ($\aleph_0$). The proof of this apparent paradox is very simple and clever, but lies far outside the scope of this book.[10] Let us instead go back to Borges's notes. On the first page of William James's *Some Problems of Philosophy*, a note by Borges ("paradox of the numerosity of numbers") refers to a passage where James describes transfinite numbers and Cantor's theory. Another book by Bertrand Russell, *Introduction to Mathematical Philosophy*, has a note on the last page, "parts that resemble the whole, p. 81," referring to similar descriptions.

Borges had more than a dozen books on mathematics in his library (many of them, curiously, dealing with the fourth dimension, a topic about which he seems to have been passionate).[11] In his preface to

10. Guillermo Martínez offers a lucid and intuitive proof in his book *Borges y la matemática* (Buenos Aires: Seix Barral, 2003).

11. Borges discusses the fourth dimension in the essay "Un resumen de las doctrinas de Einstein" (A Summary of Einstein's Doctrines), which appeared in volume 4 of his *Obras completas* (Buenos Aires: Emecé, 2002).

a Spanish translation of Kasner and Newman's *Mathematics and the Imagination*[12] he briefly addresses the topic of infinity, and in a review of Eric Temple Bell's *Men of Mathematics* he makes an intriguing comparison between Zeno of Elea and Georg Cantor, arguing that twenty-three centuries later "the German's strange transfinite numbers were conceived to somehow solve the Greek's riddles."[13] I do not stop to ponder the mathematical validity of such a statement but simply take in and enjoy the beauty of the idea.

These notes give an idea of what Borges read and knew about mathematics and infinity. But let us go back to the subject of this book, from the infinity of the Aleph to the infinity of memory. What is there in Borges's library about Funes, memory, and the workings of the brain?

Borges had several books by William James, and although they are about philosophy—since James, apart from having set out with astounding clarity many of the current principles of neuroscience, was a brilliant philosopher—they also discuss at length different aspects of how memory and the brain work. Perhaps the most direct reference to Funes is to be found in *The Mind of Man*, a 1902 book by English psychologist Gustav Spiller. A note by Borges on the first page, "memories of a whole life, page 187,"

12. In Edward Kasner and James R. Newman, *Matemáticas e imaginación*, trans. José Celdeiro Ricoy (Buenos Aires: Consejo Nacional para la Cultura y las Artes, 2007), p. 11. (Original English edition, *Mathematics and the Imagination* [New York: Simon and Schuster, 1940].)

13. Jorge Luis Borges, review of *Men of Mathematics* by E. T. Bell, in *Obras completas*, vol. 4, pp. 461–462.

refers to a passage in which Spiller describes the number of memories a person has in the different stages of life: about 100 during the first ten years, some 3,600 up to age 20, an additional 2,000 between 20 and 25, eventually reaching about 10,000 at age 35. To illustrate his method of quantifying the number of memories (shown in parentheses), Spiller considers a trip he made from Torquay to Totnes:

When at Torquay I went (1) to Tottness [sic] (2) by rail; (3) I have a glimpse of my arrival (4) and another of part of my walk; (5) I see the pier; (6) we are going down the Dart by boat; (7) now the river seems to become narrow; (8) now I see the hills; (9) the river is frequently narrow; (10) I have pointed out to me (11) a barn.

On page 187 the author also describes the time it takes to compile these remembrances:

A somehow Bohemian life is thus numerically summed up in what may be lived through in half-a-day. In other words, I am able to re-develop about one 10,000th of what happened to me, though it must be admitted that a quantitative statement is not wholly satisfactory. Moreover, the re-developed systems of the first nine years do not cover more than the space of a minute's activity, and could be re-developed during that short interval; and if we apportion ten hours of waking existence to a day, I do not produce one 250,000th part of what happened to me during the above-mentioned period.[14]

The relation with Funes is striking; says Borges about Funes:

14. Gustav Spiller, *The Mind of Man: A Text-book of Psychology* (London: S. Sonnenschein, 1902), p. 187.

Podía reconstruir todos los sueños, todos los entresueños. Dos o tres veces
había reconstruido un día entero; no había dudado nunca, pero cada recon-
strucción había requerido un día entero.[15]

[He could reconstruct all dreams and all half-dreams. Two or three times
he had reconstructed an entire day; he had never fumbled, but each reconstruction had taken an entire day.]

Like Borges, William James argues in his *Principles of Psychology*
that a perfect memory would connote the problem that remembering any particular fact or deed would take as much time as the fact
or deed itself. This would take far too long, and we would be incapable of abstracting the fundamental content of each event in order
to develop our thought. Every recollection is achieved through a
great simplification, in the sense that we omit an enormous amount
of detail. Writes James, quoting Théodule Ribot:

"As fast as the present enters into the past, our states of consciousness
disappear and are obliterated. Passed in review at a few days' distance,
nothing or little of them remains: most of them have made shipwreck in
that great nonentity from which they never more will emerge, and they have
carried with them the quantity of duration which was inherent in their
being."[16]

A note by Borges on an 1898 English edition of John Stuart Mill's
A System of Logic, "against the language invented by Ireneo Funes,

15. Jorge Luis Borges, "Funes el memorioso," in *Obras completas*, vol. 1,
pp. 583–590.

16. William James, *The Principles of Psychology*, authorized ed., vol. 1
(New York: Henry Holt, 1890; repr., New York: Dover, 1950), p. 680.

436," gives another clear reference. In this passage, Mill highlights the need to generalize and categorize:

Even if there were a name for every individual object, we should require general names as much as we now do. Without them we could not express the result of a single comparison, nor record any one of the uniformities existing in nature; and should be hardly better off in respect to Induction than if we had no names at all. . . . It is only by means of general names that we can convey any information, predicate any attribute, even of an individual, much more of a class.[17]

In "Funes the Memorious," Borges remarks that John Locke once imagined an impossible language in which every individual thing (every stone, every bird, every branch) would have its own name:

Funes proyectó alguna vez un idioma análogo, pero lo desechó por parecerle demasiado general, demasiado ambiguo. En efecto, Funes no sólo recordaba cada hoja de cada árbol de cada monte, sino cada una de las veces que la había percibido o imaginado.

[Funes once envisioned an analogous language but abandoned the project because it seemed too general, too ambiguous to him. Indeed, Funes remembered not only every leaf on every tree on every mountain, but also each time he had perceived or imagined it.]

This lack of categorization, of logical arrangement, also led him to devise a peculiar numbering system:

17. John Stuart Mill, *System of Logic Ratiocinative and Inductive* (London: Longmans, Green, 1886), p. 436.

En lugar de siete mil trece, decía (por ejemplo) Máximo Pérez; en lugar de siete mil catorce, el ferrocarril; otros números eran Luis Melián Lafinur, Olimar, azufre, los bastos, la ballena, el gas, la caldera, Napoleón, Agustín de Vedia. En lugar de quinientos, decía nueve.[18]

[Instead of, say, seven thousand thirteen, he would say Máximo Pérez; instead of seven thousand fourteen, the railroad; other numbers were Luis Melián Lafinur, Olimar, sulfur, the suit of clubs, the whale, gas, the cauldron, Napoleon, Agustín de Vedia. Instead of five hundred, he would say nine.]

Following Mill and James, Borges makes a fanciful argument about the importance of generalization, categorization, and the generation of concepts. The capacity for abstraction in the use of language or ideas—a quality absent in Funes—is indeed a basic component of the ability to think. In "The Analytic Language of John Wilkins" (from *Other Inquisitions*, 1952), Borges gives a hilarious description of the sterility of a senseless categorization by citing from the (apocryphal, of course) Chinese encyclopedia *A Heavenly Emporium of Benevolent Knowledge*:

En sus remotas páginas está escrito que los animales se dividen en a) pertenecientes al Emperador, b) embalsamados, c) amaestrados, d) lechones, e) sirenas, f) fabulosos, g) perros sueltos, h) incluidos en esta clasificación, i) que se agitan como locos, j) innumerables, k) dibujados con

---

18. I cannot help noticing that some of these attributes resemble the jargon of the Quiniela, one of Argentina's most popular betting games, whose results are summarized using the last two digits of each number through time-honored denominations: 77, say, is "a woman's legs," 29 is "Saint Peter," and 48 is "the dead man that speaks."

un pincel finísimo de pelo de camello, l) etcétera, m) que acaban de romper el jarrón, n) que de lejos parecen moscas.[19]

[In its remote pages it is written that animals can be classified as a) the Emperor's, b) embalmed, c) tame, d) suckling, e) mermaids, f) imaginary, g) stray dogs, h) included in this classification, i) that shake like madmen, j) innumerable, k) drawn with a very fine brush of camel hair, l) etcetera, m) that just broke the vase, n) that from afar resemble flies.]

19. Jorge Luis Borges, "El idioma analítico de John Wilkins," in *Obras completas*, vol. 2, pp. 102–106. I would like to thank Mariano Sigman, professor in the Department of Physics of the Faculty of Exact Sciences of the University of Buenos Aires, for pointing out the relationship between Wilkins's language and Funes's memory. Part of Mariano's research is on language and the similarity between concepts and words, a topic closely related to Wilkins's language.

THE MAN WHO COULD NOT FORGET

Alexander Romanovich Luria (1902–1977) was a brilliant Russian psychologist who, along with Alexei Leontiev and Lev Vygotsky, created a new school of thought at the Institute of Psychology of the University of Moscow. These three scientists (who called themselves the *troika*) set out to reconcile two opposing approaches: on one side were the followers of Wilhelm Wundt, who considered psychology a natural science and spent their time in the laboratory carrying out experiments with the purpose of reducing complex psychological processes to fundamental mechanisms (this school became dominant in Russia following Pavlov and his study of conditioned reflexes); on the other were those that followed Wilhelm Dilthey in seeing psychology as a humanist discipline, more descriptive and phenomenological, and believed that laboratory experiments were incapable of explaining complex mechanisms such as those that generate consciousness (Wundt's followers argued that those processes could not be studied scientifically).[1]

Luria obtained his psychology degree from the University of Kazan at the age of 19 and further pursued his interest in brain pathologies as he worked toward a medical degree that he earned a few years later. This multifaceted education, both scientific and clinical, put him in a privileged position to contribute an approach that would supersede the famous dichotomy between Wundt's experimental psychology, which was rather removed from reality, and Dilthey's phenomenology, which lacked scientific rigor. Among the books published by Luria is *The Man with a Shattered World*, in which he described Zasetsky, a patient who had suffered a cranial trauma during the Second World War (when a bullet penetrated his parieto-occipital cortex) and who, among other problems, perceived objects in a fragmented way and had problems with reading and writing. In another book, *The Mind of a Mnemonist*, Luria chronicles the case of Solomon Shereshevskii, who possessed an extraordinary memory. One of Luria's most notable traits was his ability to portray the human side of his patients without sacrificing

1. Nowadays, most scientists (myself included) have a point of view diametrically opposed to Wundt's and, following guidelines set forth by Francis Crick and Christof Koch, study the neural correlates of consciousness. In a recent report we demonstrated that the neurons in a region of the brain called the hippocampus are activated only when the subject consciously perceives a visual stimulus (and remain silent when the same stimulus is not recognized). For details, see Rodrigo Quian Quiroga, Roy Mukamel, Eve Isham, Rafael Malach, and Itzhak Fried, "Human Single Neuron Responses at the Threshold of Conscious Recognition," *Proceedings of the National Academy of Sciences of the USA* 105 (2008): 3599–3604.

scientific rigor, somehow linking Wundt's and Dilthey's points of view; he was interested not only in the fascinating details of their pathologies but also in their personalities and quality of life. In a scientific world that praises precision and objectivity, Luria brought a breath of fresh air through what came to be known as "romantic science," a style later taken up by neurologist Oliver Sacks in books like *Awakenings*[2] and *The Man Who Mistook His Wife for a Hat*.

It is precisely from the description of Shereshevskii as a person, transcending the merely scientific, that notable parallels between him and Funes emerge—in Luria's case, through a groundbreaking approach to studying his patients; in Borges's, exclusively as a product of his imagination.

Solomon Shereshevskii (whom, following Luria, we shall also call S.) worked as a journalist for a Moscow newspaper. Every morning the editor would distribute among the staff a long list of assignments that included addresses of places, information to be obtained, etc., and one day, when he noticed that S. took no notes, he chided him for apparently not paying attention. However, to the editor's surprise, S. recited word for word his tasks for the day. The editor then suggested that S. visit Luria, who at the time was a very young psychologist in the early stages of his career. Already on their first encounter in the 1920s S. surprised Luria by being able to repeat effortlessly sequences of 30, 50, and even 70 letters or numbers; moreover, he could repeat the sequences in reverse order, or call out the item that preceded or followed any part of the sequence that

2. This book was made into a movie of the same name starring Robert De Niro and Robin Williams.

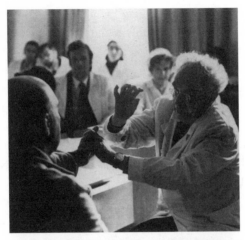

FIGURE 3.1
Top: Alexander Luria in the 1960s, examining a patient. Bottom: Solomon Shereshevskii. (I am indebted to Sergey Antopolsky for finding this very rare picture of Shereshevskii.)

Luria named. It made no difference to S. whether the sequence was composed of words, random syllables, numbers, or sounds—he would simply visualize the sequences in his memory and had only to read them back from this visualization. Luria in fact acknowledged having failed at perhaps the easiest task for a psychologist: measuring someone's memory capacity.

Luria verified that S. kept these sequences in his memory for many years after having learned them, even though he did not expect ever to be asked about them again. Some examples are incredible. For example, Luria once read to S. the first four lines of Dante's *Divine Comedy*:[3]

Nel mezzo del cammin di nostra vita
mi ritrovai per una selva oscura,
ché la diritta via era smarrita.
Ahi quanto a dir qual era è cosa dura . . .

Not only did S. recite these verses (and several subsequent stanzas) with the correct pronunciation despite not knowing Italian,[4] but he did it again during another encounter with Luria 15 years later. On another occasion, Luria asked S. to remember a made-up, meaningless mathematical formula:

$$N \cdot \sqrt{d^2 \times \frac{85}{vx}} \cdot \sqrt[3]{\frac{276^2 \cdot 86x}{n^2 v \cdot \pi 264}} \ n^2 b = sv \frac{1624}{32^2} \cdot r^2 s$$

3. Incidentally, this book was beloved by Borges, who had several editions and translations of it in his library.

4. Given the differences between Italian and Russian, to S. these verses were just unknown, meaningless words.

S. repeated the formula without any mistake and reconstructed it 15 years later (again, not knowing that Luria would ask him about it once more).

Another interesting situation arose when S. was asked to remember the following sequence of syllables:

1.	ma	va	na	sa	na	va
2.	na	sa	na	ma	va	
3.	sa	na	ma	va	na	
4.	va	sa	na	va	na	ma
5.	na	va	na	va	sa	ma
6.	na	ma	sa	ma	va	na
7.	sa	ma	sa	va	na	
8.	na	sa	ma	va	ma	na

etc.

This sequence was one of the most difficult, as it is composed of four meaningless syllables that are easy to mix up given their similarity to each other. S. managed nonetheless to memorize the sequence and to repeat it four years later.

S.'s incredible memory was based on a very strong synesthesia. People with synesthesia can usually mix perceptions from different senses, for example associating numbers with colors. In S.'s case, however, these associations went much further: every letter, number, or word set off a downpour of visual imagery, sounds, tastes, and tactile sensations. In the course of the 30 years he spent studying S., Luria took a large number of notes about S.'s experiences and his unique perception of reality. In the rest of this chapter I will transcribe some of Luria's notes which help give an idea of how it is to live with such an extraordinary memory. During a visit in 1939, S. told Luria:

I recognize a word not only by the images it evokes but by a whole complex of feelings that image arouses. It's hard to express . . . it's not a matter of vision or hearing but some over-all sense I get. Usually I experience a word's taste and weight, and I don't have to make an effort to remember it—the word seems to recall itself.[5]

These words were indelibly engraved in his memory due to their high visual content.[6] As S. says during a session in 1936:

When I hear the word *green*, a green flowerpot appears; with the word *red* I see a man in a red shirt coming toward me; as for *blue*, this means an image of someone waving a small blue flag from a window. . . . Even numbers remind me of images. Take the number 1. This is a proud, well-built man; 2 is a high-spirited woman; 3 a gloomy person (why, I don't know); 6 a man with a swollen foot; 7 a man with a mustache; 8 a very stout woman—a sack within a sack. As for the number 87, what I see is a fat woman and a man twirling his mustache.[7]

These associations were not unique, and could be very rich in that they involved different senses. In another session with Luria, S. says:

5. Alexander Luria, *The Mind of a Mnemonist: A Little Book about a Vast Memory*, trans. Lynn Solotaroff (Cambridge, MA: Harvard University Press, 1987), p. 28. I have taken the liberty to reproduce these notes at length because they are not likely to be well known among nonspecialist readers.

6. It is worth noting that most of the human brain is devoted to processing visual stimuli. Thus associating words with visual cues is a very efficient way to exploit the machinery of the brain.

7. Luria, *The Mind of a Mnemonist*, p. 31.

For me 2, 4, 6, 5 are not just numbers. They have forms. 1 is a pointed number—which has nothing to do with the way it's written. It's because it's somehow firm and complete. 2 is flatter, rectangular, whitish in color, sometimes almost a gray. 3 is a pointed segment which rotates. 4 is also square and dull; it looks like 2 but has more substance to it, it's thicker. 5 is absolutely complete and takes the form of a cone or a tower—something substantial. 6, the first number after 5, has a whitish hue; 8 somehow has a naïve quality, it's milky blue like lime.[8]

Such was S.'s degree of synesthesia that Luria's notes seem to be describing a lunatic. These notes do a masterful job of describing S.'s mind and the way he could firmly fix memories in his brain starting from richly detailed associations—associations that look bizarre to us but were perfectly natural to him. While for Ireneo Funes it was natural to remember everything, the secret of S.'s extraordinary memory lay in his synesthesia.

Luria also investigated the way in which S. could memorize long sequences of words. The trick was simple. Each word would spark an image that he would distribute along a path he had devised in his mind.[9] For that, he would use a street in his hometown or a

8. Ibid., p. 26.

9. This technique is known as the "method of loci" (loci meaning "places" in Latin). It was developed by Simonides of Ceos (556–468 B.C.) as a result of a dramatic circumstance: a roof collapsed during a banquet and killed all of the guests except Simonides, who had just stepped out to receive a message. The victims' bodies were so disfigured that they could only be identified thanks to Simonides, who remembered where everyone had been seated. As a result, Simonides realized the importance of ordering facts in order to keep them in memory and invented mnemonics. Techniques like the method of loci were of great importance back then, when

famous street in Moscow. S. would then stroll along that street in his mind, visualizing all the elements—the words—he had left there. This technique allowed him to repeat the sequence forward or backward, or recall the word that preceded or followed another word. Once he had formed an image of the street in question, S. had only to wander about and name whatever he saw.

Perhaps the most fascinating twist in Luria's study of S. came when, frustrated at his inability to find a limit to S.'s memory, he decided to deal with a less obvious but much more interesting question: Could S. forget?

Generally, whenever S. made a mistake it was not due to a memory lapse but to a perception problem: he had either failed to pay enough attention to a particular word, or he had left its visual representation in an obscure place during his imaginary stroll. For example, during a session in 1932 he justified a mistake by saying the following:

I put the image of the *pencil* near a fence. . . . But what happened was that the image fused with that of the fence and I walked right on past without noticing it. . . . Sometimes I put a word in a dark place and have trouble seeing it as I go by. Take the word *box*, for example. I'd put it in a niche in the gate. Since it was dark there I couldn't see it.[10]

Incredibly, S.'s problem was not how to remember, but how to forget. This became much worse when he started to work as a

orators—be they politicians, philosophers, or artists—would give their speeches exclusively from memory.

10. Luria, *The Mind of a Mnemonist*, p. 36.

professional mnemonist. Several times a night, audience members would give S. long sequences to remember, and these started to pile up in his memory at a frenetic pace. Like Funes, who could remember every single detail and wound up locking himself in a dark room, S. was unable to forget things that were no longer relevant, and these memories became a torment.

To give an idea of the complexity of the problem, so far removed from a normal person's experience, I transcribe S.'s often tragic account of the methods he used for trying to forget. Everything he was supposed to remember during a show was written on a blackboard and erased at the end. In 1939, he said to Luria:

I'm afraid I may begin to confuse the individual performances. So in my mind I erase the blackboard and cover it, as it were, with a film that's completely opaque and impenetrable. . . . Even so, when the next performance starts and I walk over to that blackboard, the numbers I had erased are liable to turn up again. If they alternate in a way that's even vaguely like the order in one of the previous performances, I might not catch myself in time and would read off the chart of numbers that had been written there before.[11]

S. also tried writing everything that he wanted to forget. He reasoned that he would not have to remember something once it had been put in writing, just as one feels free to forget a telephone number after writing it in an address book. This method did not work, though, since he continued to see what he had written. He even tried, again fruitlessly, to throw away or burn the sheets where he had written his memories, as though the physical act of

11. Ibid., p. 69.

destroying the notes could erase them from his head. Finally the day came when he realized that the only way to erase unwanted memories was to elude them deliberately. All he had to do was to avoid paying attention to such memories—for example, the particular sequences that he had memorized during a performance. And this takes us back to the apparent paradox described by Luria: S. had to make an effort to forget.

To relieve himself of these memories, S. also employed the "eido-technique": instead of remembering every detail of an image, he would pick just one that encapsulated the whole meaning. This way he could compress images and abstract their main features, like when we generalize a concept. S. had to make a voluntary effort to achieve what is innate in everyone else: to forget details and generalize. Luria explains that S.'s childhood memories were far richer than those of a normal person because they were never transformed into concepts or words (in adults, visual images are replaced by ideas that relate to their meaning). According to Luria, primal visual thought gives way to one that is more verbal and logical, in which visual images are kept in the periphery of consciousness because they do not help to understand abstract concepts.

S.'s struggles to forget bring us to the most interesting part of Luria's narrative, where he describes the drawbacks of having such an extraordinary memory. Luria writes that S. was quite inept at anything having to do with logical reasoning. His memory worked exclusively through visual imagery and employed no logic when memorizing. For example, during a session with Vygotsky and Leontiev (Luria's colleagues, whom we mentioned earlier in this chapter) he was asked to memorize a sequence of words that included names of birds and later another that included names of different liquids. Though he could repeat both sequences, S. was incapable of naming the birds in the first sequence or the liquids in the second. But perhaps the most interesting example of S.'s logical deficiency came when he was asked to memorize this list of numbers:

```
1   2   3   4
2   3   4   5
3   4   5   6
4   5   6   7
etc.
```

S. remembered the sequence using the amazing power of his visual memory but failed to notice that the numbers were consecutive— that is, they followed a logical rule that made them very easy to remember.

S.'s lack of skill with any sort of abstract and nonvisual thought also manifested itself as an inability to grasp the content of whatever he read. Though capable of reciting by rote the first few lines of the *Divine Comedy*, he was unable to pinpoint the content of a book to understand its meaning and follow its narrative. In other words, while ordinary people synopsize and remember a few facts that let them follow a story as it develops, it took Luria almost three pages to describe the associations that S. made as he learned the first four lines of the *Divine Comedy*. Though these associations allowed him to recite the text from memory, he had to fight against an incontrollable avalanche of images and associations sparked by each and every word whenever he attempted to concentrate on the essential information communicated by the text. During an encounter in 1937, S. told Luria:

It's particularly hard if there are some details in a passage I happen to have read elsewhere. I find then that I start in one place and end up in another—everything gets muddled. Take the time I was reading *The Old World Landowners*. Afanasy Ivanovich went out on the porch. . . .Well, of course, it's such a high porch, has such creaking benches. . . . But, you know, I'd already come across that same porch before! It's Korobochka's porch, where Chichikov drove up! What's liable to happen with my images is that Afanasy Ivanovich could easily run into Chichikov and Korobochka![12]

In another session, S. confessed to Luria that he could not read nor study because these unwanted associations made him lose track of what he was reading. Poetry, in particular, demands a figurative understanding, the ability to grasp the ideas suggested by words so as to go beyond their literal meaning and the precise images they evoke by themselves. It is therefore not surprising that S. considered poetry to be a nightmare.

This limitation in S.'s ability to abstract had even graver consequences. For example, it was difficult for him to recognize voices. As he told Luria in 1951,

I got so interested in [filmmaker Sergei M. Eisenstein's] voice, I couldn't follow what he was saying. . . . But there are people whose voices change constantly. I frequently have trouble recognizing someone's voice over the phone, and it isn't merely because of a bad connection. It's because the

12. Ibid., p. 113. Luria's translator includes an illuminating footnote at the end of this passage: "The characters [S.] describes are from Gogol's *Dead Souls* and some of the stories in his Ukrainian tales. S.'s reading leads to a state of confusion in which characters from the different works come together in a single image."

person happens to be someone whose voice changes twenty to thirty times in the course of a day. People don't notice this, but I do.[13]

During another session, S. told Luria that he ran into trouble when he tried to recognize faces because they change all the time:

A person's expression depends on his mood and on the circumstances under which you happen to meet him. People's faces are constantly changing; it's the different shades of expression that confuse me and make it so hard to remember faces.[14]

Compare this to what Borges says about Funes:

No sólo le costaba comprender que el símbolo genérico perro abarcara tantos individuos dispares de diversos tamaños y diversa forma; le molestaba que el perro de las tres y catorce (visto de perfil) tuviera el mismo nombre que el perro de las tres y cuarto (visto de frente). Su propia cara en el espejo, sus propias manos, lo sorprendían cada vez.[15]

[Not only was it difficult for him to understand that the generic term "dog" could embrace so many disparate individuals of diverse sizes and shapes; it bothered him that the dog seen in profile at 3:14 would be called the same as the dog at 3:15 seen from the front. His own face in the mirror, his own hands, surprised him every time.]

13. Ibid., pp. 24–25.

14. Ibid., p. 64.

15. Jorge Luis Borges, "Funes el memorioso," in *Obras completas* (Buenos Aires: Emecé, 2007), vol. 1, pp. 583–590.

LIVING IN THE PAST

One of the most spectacular contributions to our understanding of how memory works comes from the study of Henry Gustav Molaison (1926–2008), who, until his death and for privacy reasons, was known as Patient H.M.

Molaison started to suffer from epileptic seizures at age 10, probably as a consequence of a serious blow to the head (sustained when he was run over by a bicycle) that made him lose consciousness for several minutes. The seizures worsened during adolescence and at some point could not be treated with medication, no matter how high the dose. As happens to other people who suffer from intractable epilepsy, H.M.'s quality of life deteriorated greatly, and at age 27 he had to quit his job at a motor factory. At that point, and as a last resort, neurosurgeon William Scoville proposed that Molaison undergo an experimental surgical procedure based on Scoville's

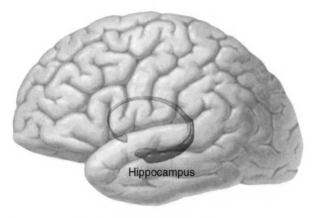

Hippocampus

FIGURE 4.1
The hippocampus is located some two centimeters (about an inch) into the brain, at about the same height as the ears.

experience with psychotic patients, who had shown significant improvement following the removal of part of the frontal lobe. Molaison and his family consented, and in 1953 Scoville surgically removed Molaison's hippocampi—the seahorse-shaped structures that are often related to the onset of epileptic seizures—and adjacent zones on both hemispheres of his brain. The surgery, which indeed managed to stop his seizures, radically changed the history of neuroscience and our knowledge of memory, but unfortunately transformed Henry Molaison into Patient H.M. forever.

Following the surgery, H.M. underwent what at first appeared to be a normal recovery, but soon thereafter a terrible and unexpected shortcoming became evident: H.M. could not recognize the staff at the hospital, find his way to the bathroom, or remember daily events. In brief, H.M. was incapable of forming new memories, a condition known as anterograde amnesia (in contrast with the better-known retrograde amnesia, in which a person forgets past memories, for example after sustaining a blow to the head). In a scientific paper describing H.M.'s case, Scoville and his colleague Brenda Milner[1] report that, several months after moving with his family to a new home, H.M. was unable to remember his new address but had no problem recalling the previous one perfectly. Neither could he remember where he kept his everyday objects at home. He could solve the same puzzles again and again without ever learning them, and could read the same magazine several times without recognizing its content. One of his favorite pastimes was tending the garden, but his mother had to tell him every day where the lawnmower was, even if he had used it the day before. H.M. could not remember what he had had for lunch just half an hour after the meal; in fact, he could not even remember having had lunch at all, and was as hungry and thirsty after a meal as before it. H.M. lived in the past and in a fleeting present that continually, inexorably, faded to oblivion.

An extraordinary description of anterograde amnesia can be seen in Leonard Shelby (Guy Pearce), the main character in Christopher Nolan's film *Memento*, who in order to remember something relating to someone had to scribble a brief note on a Polaroid of that person. The movie revolves around some dramatic decisions that Leonard has to make based on these notes, as though the notes were his only source of remembrance. One particularly harrowing

---

1. William Scoville and Brenda Milner, "Loss of Recent Memory after Bilateral Hippocampal Lesion," *Journal of Neurology, Neurosurgery, and Psychiatry* 20 (1957): 11–21.

FIGURE 4.2
Henry Molaison (H.M.) before undergoing surgery and, years later, in his garden.

scene shows Leonard's desperation when he is unable to find a pen to scrawl an important note on a photo before the memory fades away for good. Curiously, in this case the pen plays the role of the hippocampus: it turns existing information into long-term memories.

After seeing the result of H.M.'s surgery, Scoville dedicated him-self—mainly through scientific papers and conferences—to pre-venting others from making the same mistake that he had unwittingly made: surgically removing the hippocampi from both brain hemi-spheres in epileptic patients. Scoville actually succeeded in prevent-ing the surgery from being performed ever again, turning H.M. into a unique case in the history of neurology. Nowadays, the removal of one of the hippocampi (never both) from patients whose epilepsy cannot be controlled with medication is a very successful procedure.

In fact, most patients who undergo this surgery[2] are cured of their epilepsy with virtually no collateral damage, since the hippocampus that stays in the brain tends to take over the function of the one that has been removed. This type of procedure occasionally grants us access to the behavior of the human brain through intracranial recordings, as we shall see in subsequent chapters.

Wilder Penfield, a distinguished neurosurgeon, together with his student Brenda Milner, a psychologist, presented two cases similar to H.M.'s (though featuring a less extreme memory loss)[3] at the 1955 annual meeting of the American Neurological Association. When he found out about these two patients, Scoville phoned Penfield, and thus, almost serendipitously, Milner ended up studying H.M. for 30 years—interestingly, the same number of years that Luria devoted to Shereshevskii.[4] However, and in spite of their

2. Unfortunately, not all patients can be subjected to this type of surgery: some of them have seizures that originate in vital areas that cannot be surgically removed, while others have seizures whose focus is not circumscribed to a particular area.

3. The difference from H.M.'s case is that Penfield removed only one of the hippocampi in these patients. Penfield and Milner postulated—before they found out about H.M.—that their patients' observed memory deficit might have been due to a bilateral lesion, since the other hippocampus might have been damaged by epilepsy. This was verified some 10 years later by means of a postmortem autopsy.

4. Upon finding out about her discoveries with H.M., Luria wrote Milner: "Memory was the sleeping beauty of the brain, and now she is awake." (Quoted from the radio program *All in the Mind* on ABC Radio National, Australia, May 28, 2011.)

almost quotidian interaction, every one of Milner's encounters with H.M. was like the first. She had to introduce herself to H.M. every single time, as though she were a complete stranger.

During a psychological test in April of 1955 (a year and a half after the surgery), H.M. estimated the date to be March of 1953 and stated that he was 27 years old, when he had already turned 29. He was barely aware of having had surgery. Several tests showed that H.M. had no problems with visual perception or general reasoning (as long as these required no memory use). In fact, an intelligence test gave him an above-average IQ of 112. Inevitably, however, as he embarked upon any new test he was unable to remember whatever he had done just before. He was unable even to recognize the test he had just taken if it was given to him again.

Fourteen years after the surgery, Brenda Milner and her student Suzanne Corkin—who also went on to study H.M. for years— reported that he was still incapable of recognizing the people he had met after the surgery—new neighbors, new family friends, and even Milner and Corkin themselves.[5] They found that H.M.'s amnesia was not limited to what happened after the surgery but also comprised a few previous years. Despite his persistent memory deficit, H.M. understood language like a normal person, was able to hold a conversation, and could understand jokes, even those based on double meanings (which require some capacity for reasoning). This showed that his short-term memory was working adequately, since without it it is impossible to form sentences, speak coherently, or

5. Brenda Milner, Suzanne Corkin, and H.-L. Teuber, "Further Analysis of the Hippocampal Amnesic Syndrome: 14-Year Follow-Up Study of H.M.," *Neuropsychologia* 6 (1968): 215–234.

understand what someone else is saying. Milner describes H.M. as a person who is always about to understand his circumstances but cannot completely grasp what is going on because he cannot place his current situation within the context of his immediate past. H.M. said his life was like waking from a dream. Most days we wake up on our beds, in familiar surroundings, but sometimes, for example if we spend the night at a hotel far away from home, upon waking up we have a brief period of confusion in which we don't know where we are or what we are doing there. Then our memory quickly rescues us, reminding us why we happen to be where we are, and everything makes sense again. (Curiously, Borges repeatedly talked about using the word "remembering" instead of "waking," since, in his view, one has no memory of oneself while asleep, and this memory comes back only upon awakening.)[6] In H.M.'s case, this moment of confusion extended indefinitely because he had no memory of where he was and what he was doing, wherever he happened to be. The first scene of *Memento* starts with Leonard waking up at a place he cannot recognize, talking to himself:

So, where are you? You're in some motel room. You just wake up and you're in . . . in a motel room. There's the key. It feels like maybe it's just the first time you've been there but . . . perhaps you've been there for a week . . . three months. . . . It's kind of hard to say . . . I don't know, it's just an anonymous room.

As the years went by, H.M. started to come to terms with the fact that he was at a hospital, but was still unable to remember a single

6. *Jorge Luis Borges: Conversations*, ed. Richard Burgin (Jackson: University Press of Mississippi, 1998), p. 166.

specific detail of what had occurred in previous days. He would say, "Every day is alone in itself, whatever enjoyment I have had and whatever sorrow I have had."[7]

H.M. was always willing to participate in experiments, even shortly before his death. He was studied by more than 100 scientists, though mainly by Brenda Milner and Suzanne Corkin, and the results of those investigations were published in several scientific journals. We can summarize the most significant discoveries by stating that H.M. had no deficiencies in perception, intelligence, or language processing. He also had no short-term memory problems. He could repeat sequences of six or seven numbers and remember something for some 20 to 30 seconds, though the only way he had to extend those memories was through constant repetition; he could even remember a three-digit number for 15 minutes, so long as he was not distracted.

H.M. could remember things that preceded the surgery but was unable to generate new memories. He did not know the meaning of new words. His memory deficit made him require constant care: he had to be reminded to shave, brush his teeth, comb his hair, eat. As time went by he became aware of his problem. Despite not being able to remember certain facts, after many repetitions he familiarized himself with new situations—or perhaps with the mere fact of being permanently disoriented. For example, at first he would call a nurse in the middle of the night to ask her where he was and why, until he finally started to become familiar with his surroundings

7. Milner, Corkin, and Teuber, "Further Analysis of the Hippocampal Amnesic Syndrome."

and could find his way to his room in the assisted-living facility where he lived for years.

Borges mentioned his fear of mirrors in several instances. He begins his poem "Mirrors" (from *The Maker*, 1960) thus:

Yo que sentí el horror de los espejos
no sólo ante el cristal impenetrable
donde acaba y empieza, inhabitable,
un imposible espacio de reflejos . . .[8]

[I, who felt the horror of the mirrors,
not only as I faced the impenetrable crystal
where, uninhabitable, ends and begins
an impossible space of reflections . . .]

and writes this in "The Mirror" (from *A History of Night*, 1977):

Yo, de niño, temía que el espejo
me mostrara otra cara o una ciega
máscara impersonal que ocultaría
algo sin duda atroz. Temí asimismo
que el silencioso tiempo del espejo
se desviara del curso cotidiano
de las horas del hombre y hospedara
en su vago confín imaginario
seres y formas y colores nuevos.[9]

8. Jorge Luis Borges, "Los espejos," in *Obras completas* (Buenos Aires: Emecé, 2007), vol. 2, pp. 228–229.

9. Jorge Luis Borges, "El espejo," in *Obras completas*, vol. 3, p. 222.

[As a child, I feared the mirror
would show me another face, or a blind,
impersonal mask that would be concealing
something no doubt atrocious. I also feared
that the silent time of the mirror
would deviate from the daily course
of man's hours and would house
new beings and shapes and colors
within its vague, imaginary confines.]

Who could have been more terrified at seeing himself in the mirror than H.M., whose reflected face showed the many decades he had lived since his surgery but whose mind remained frozen at 27 years of age? Suzanne Corkin relates, however, that H.M. did not seem surprised when seeing himself in the mirror (at least judging from his demeanor and tone of voice). This shows a certain degree of familiarity with reality that stands in noticeable contrast with his lack of memory. In other words, H.M. learned to live with the fact that he was older than he remembered being. Curiously, when he was 55 he failed to recognize himself in a photo taken when he was 40; the image differed from his remembered 27-year-old likeness, but also from his 55-year-old face, with which he was familiar.

Terrifying as it must have been for H.M. to see himself in the mirror as an old man while believing himself to be 27, one can only imagine the anguish caused by his inability to come to terms with his parents' death—by having to face the sad truth again and again, always as though for the first time. His father died in 1967 and his mother four years later. Two months after his father's passing, H.M. already had a vague idea of that absence. Something similar happened with his mother: he could not remember the circumstances of her death but, with time, got used to her not being around. It is worth mentioning that, according to current scientific belief, the brain circuits that encode familiarity may be different from those that take part in recognition.[10] Familiarity—the feeling of knowing someone without actually remembering who it is—may be codified by neurons in the perirhinal cortex, an area close to the hippocampus that was not completely removed during H.M.'s surgery, while recognition—remembering the person's name, the place where we have met, etc.—may be codified in the hippocampus itself.

The surgery did not affect H.M.'s intellect nor his personality; he continued to be an affable, good-humored person. After the operation H.M. lived with his parents, then with a relative with experience in convalescent care, and finally at an institution. He kept on performing such mundane tasks as mowing the lawn, cooking, making his bed, watching television, playing bingo, and solving crossword puzzles, but was unable to work or carry out more complex duties. Some fragments of H.M.'s conversations with Milner and Corkin help give an idea of the devastating effects that the removal of both hippocampi had on his memory:

10. There is still no general consensus about this point. For accounts of two clashing views, see:
• H. Eichenbaum, A. P. Yonelinas, and C. Ranganath, "The Medial Temporal Lobe and Recognition Memory," *Annual Review of Neuroscience* 30 (2007): 123–152;
• Larry R. Squire, John T. Wixted, and Robert E. Clark, "Recognition Memory and the Medial Temporal Lobe: A New Perspective," *Nature Reviews Neuroscience* 8 (2007): 872–883.

Milner   When you're not at MIT, what do you do during a typical day?

H.M.   See, that's what I don't—I don't remember things.

[. . .]

Milner   Who is the president of the United States now?

H.M.   That I don't—I couldn't tell you. I don't remember exactly at all.

Milner   Is it a man or a woman?

H.M.   I think it's a man.

Milner   His initials are G.B. [for George Bush]. Does that help?

H.M.   No, it doesn't help.

[. . .]

Milner   Do you know what you did yesterday?

H.M.   No, I don't.

Milner   How about this morning?

H.M.   I don't even remember that.

Milner   Could you tell me what you had for lunch today?

H.M.   I don't know, to tell you the truth.

On the other hand, his pre-surgery memory appeared intact:

Milner   What happened in 1929?

H.M.   The stock market crashed.

[. . .]

Corkin   What do you think you'll do tomorrow?

H.M.   Whatever is beneficial.

Corkin   Good answer. Are you happy?

H.M.   Yes. Well, the way I figure it is, what they find out about me helps them to help other people.[11]

11. The conversations aired in the United States on National Public Radio's *Weekend Edition Saturday* on February 24, 2007, as part of the story "H.M.'s Brain and the History of Memory" by Brian Newhouse.

Thanks to H.M.'s case and many subsequent studies,[12] we now know with certainty that the hippocampus plays a crucial role in the establishment of new memories. The hippocampus itself is not the final storage place for (factual) memories—it codifies the information to be stored and transfers it to the cerebral cortex, which is where memories reside. The hippocampus is somehow like a clerk who compiles and archives different files and makes connections between them. Thus H.M. could remember the people he had known before he underwent surgery (the files already stored were left intact), but he was missing the "clerk" who could file away new memories.[13] This was a spectacular breakthrough for neuroscience, since, following Karl Lashley (whom we shall meet again below), it was at first believed that the system in charge of processing memories did not have a specific location within the brain.

We close for the time being our description of Henry Molaison. To give an idea of the impact of his contribution, the paper by

12. The evidence contributed by H.M. has been confirmed by studies on monkeys and on other patients whose hippocampi (and surrounding areas) had been harmed by viral encephalitis, occluded arteries, or hypoxia. In these cases, however, the injury was not as clear-cut as in H.M.'s. The results are reviewed in:
• Larry R. Squire and Stuart Zola-Morgan, "The Medial Temporal Lobe Memory System," *Science* 253 (1991): 1380–1386;
• Larry R. Squire, Craig E. L. Stark, and Robert E. Clark, "The Medial Temporal Lobe," *Annual Review of Neuroscience* 27 (2004): 279–306.

13. Another possible analogy is with the RAM in a computer. The RAM processes information, but this information is eventually sent to the hard drive, where it is finally stored.

Scoville and Milner where H.M.'s case was first described has been cited by over 2,500 subsequent papers, a stunning number considering that most scientific papers are cited a dozen times at most. The studies performed on H.M. were the first to show a link between the hippocampus and memory. As this book is being written, a simple Google search of "hippocampus and memory" yields more than 1,600,000 hits. There are over 850 journal articles that mention H.M. in the title (and, of course, many more that mention him in the text). When H.M. died, Milner admitted having a strange sensation: she felt that she had lost a close friend, albeit a friend who failed to recognize her time and again, even after decades of constant interaction.

SUBTLETIES OF MEMORY

In a series of conversations with Antonio Carrizo,[1] Borges remembers the facts and circumstances that led him to write certain lines and evokes some writers with admiration while he criticizes others with delightful wit (I cannot help recalling a fragment in which he lauds the intelligence of José Ortega y Gasset but adds that he should have hired a ghostwriter). In these interviews, as in many others, Borges quotes whole sentences—even whole paragraphs—related to the topics that Carrizo raises, be it *Martín Fierro*, Dante, Shakespeare, Verlaine, or a *milonga* that still resonates in his memory. The first, almost inevitable, question that comes to mind as one reads these dialogs is, How can he remember all that?

1. *Borges, el memorioso. Conversaciones de Jorge Luis Borges con Antonio Carrizo* (Mexico City: Fondo de Cultura Económica, 1982).

Carrizo tells Borges that he (Borges) is like Funes the memorious. Borges, however, replies that he tries not to lean too much on his memory and prefers to live forward and think about the future, because otherwise memory can become an illness or an addiction.

It is difficult to venture a scientific explanation of how Borges managed to remember so much. One can only say that literature was his passion, his life; that he could be moved to tears by a beautiful sentence in the same way that a music lover can remember in uncanny detail every single bar of an opera that stirs him. It is also possible that Borges's blindness contributed to sharpening his memory, not so much because the lack of visual stimuli made him trawl through his memories—like the character in "The Writing of the God"[2]—but because he had realized since childhood that sooner or later he would become blind and had decided he should somehow try to preserve the treasures conjured by Kipling, Chesterton, and many others.

Borges, already blind when he conversed with Carrizo, could obviously remember where his glass of water rested on the table and could also follow the conversation enough to end an exchange with one of his quips. A second question then arises, a perhaps subtler and less obvious question that nonetheless has kept scientists busy for over a century: are all these kinds of memories the

2. In this amazing short story, Borges chronicles the thoughts of Tzinacán, a Mayan shaman imprisoned by conquistadores in a stony pit from which he can see light (and a jaguar kept in the next cell whose spots resemble an inscrutable handwriting) only once a day, when his warden brings him food and water.

same? In other words: are all these memories generated by the same kind of processes in the brain? Is there a fundamental difference between my memory of how letters are distributed on my computer keyboard, which helps me to coordinate my finger movements as I type, my recall of the content of this paragraph, and the way I remember the plot of "The Writing of the God"?

The study of these subtleties of memory started with German psychologist Hermann Ebbinghaus (1850–1909), who at the end of the nineteenth century performed a series of simple, though tedious, experiments with himself as subject. Ebbinghaus constructed a set of 2,300 meaningless words, each of them composed of a vowel surrounded by two consonants (KIF, TOC, SUP, etc.), and attempted to remember random samples of between seven and thirty-six of these words. Ebbinghaus then tested his memory at different time intervals and measured the number of repetitions and the time he required to remember each list. From these experiments he inferred two fundamental principles: first, that some memories last only a few minutes, while others persist for hours, months, or years; second, that practice and repetition help memories to endure.

Ebbinghaus's first principle was elaborated further by William James,[3] who concluded that there are two types of memory: primary and secondary. Primary memory—or short-term memory, as it is currently known—allows us to keep information in our consciousness for brief periods; it is the one I use, for example, to remember the content of the sentence I am presently writing so I can finish

3. William James, *The Principles of Psychology*, authorized ed., vol. 1 (New York: Henry Holt, 1890; repr., New York: Dover, 1950).

it with the right words. This type of memory, which typically lasts only a few seconds and can store no more than about a half-dozen items,[4] is what lets us keep track of events as they flow through our present. It is the phase in which H.M. and Leonard from *Memento*—both of whom we met in the last chapter—always found themselves. As James reasoned, we do not remember what we felt when we uttered this or that word. Our awareness of such transient states is limited only to the instant of their occurrence and does not go on to become part of our past experience. Secondary—or long-term—memory, on the other hand, is the one that lets our consciousness regain the past—the one I use for example to remember Borges's features, the content of one of his stories, or a conversation about them that I had with María Kodama. As James argued, secondary memory is our knowledge of a fact or event from the past that has ceased to be part of our consciousness (because we have ceased to think about it) and that we can bring back to the present knowing that it is something we have lived through before.

Simply put, short-term memory generates our "stream of consciousness," our perception of the present. Long-term memory stores our past. Clearly, except in cases like H.M.'s or Leonard's, short-term memory can become long-term memory. I am writing these lines in a hotel, in California, and *a posteriori* I may remember some salient events about this present (like the fact that I am

---

4. In a well-known paper, George Miller argues that it is possible to store up to seven concepts in short-term memory: George Miller, "The Magical Number Seven Plus or Minus Two: Some Limits on Our Capacity for Processing Information," *Psychological Review* 63 (1956): 81–97.

about to stop writing to go to the beach). What happens then to the rest of our short-term memories? Do we forget them, just as we forget the name of some kid we knew in elementary school? No, we do not even store them in our brains. Or, as James would say, we never separate them in our consciousness from the immediate present. They are like water slipping through our fingers, something that only an imaginary being like Ireneo Funes could possibly retain.

Most short-term memories fade inevitably into oblivion, but the greatest loss of information starts even earlier, with the so-called sensory memory. This is the memory that we use, for example, to form a coherent visual image, given that each instant we see in detail only a minuscule area of our field of view (roughly the size of a thumbnail at arm's length) and our eyes continually travel back and forth over the field (executing what are called saccades) to compile the large picture. This visual, or iconic, memory is unconscious and lasts only a fraction of a second.[5] The existence of this kind of memory was demonstrated by a simple yet conclusive experiment conceived by psychologist George Sperling.[6]

Initially, Sperling asked several subjects to remember as many letters as they could from a table (like that of figure 5.1) that they could see only for a twentieth of a second. The subjects could remember between three and four letters on average. In the next

5. A similar process occurs with auditory sensory memories, also known as echoic memories, though these actually last for some three to four seconds.

6. George Sperling, "The Information Available in Brief Visual Presentation," *Psychological Monographs* 74 (1960): 1–29.

FIGURE 5.1
Example of a letter table used by Sperling to prove the existence of iconic memory.

test, however, the subjects had to repeat the letters from one row exclusively; the selected row was indicated with a low, medium, or high tone immediately after the image disappeared. Surprisingly, the subjects once more remembered between three and four letters from each line, and, since they watched the image without knowing which line they would be interrogated about, they clearly had a grasp of the whole table. This led Sperling to hypothesize the existence of a memory that precedes short-term memory and allows people to retain sensory information for extremely brief time intervals. When a person tries to recall the complete table, the sensory memory—the image of the table imprinted in the brain—fades off in the time it takes to name three or four letters; this time, however, suffices to name all of the letters in a particular row. Sperling also played the row-choosing tones with varying delays and found that the subjects' ability to recall the letters in a given row deteriorated quickly as the delay increased, thus proving that this memory lasts only for a fraction of a second.

From Sperling's experiments we can also conclude that sensory memory turns into short-term memory through attention mechanisms. Once the subjects heard the tones, they could concentrate on the line in question and forget the rest. Paying attention to a fleeting image allows us to keep it in memory for a few seconds. Now, how do we then turn some of those short-term memories into long-term memories that we may remember for years? This topic is currently the subject of intense exploration,[7] but the general scheme was laid out in Ebbinghaus's second principle, which asserts that the generation of long-term memories is facilitated through repetition. Starting from the distinction set forth by James between primary and secondary memory, and carrying out experiments similar to Ebbinghaus's, Georg Müller and Alfons Pilzecker, two German psychologists of the late nineteenth century, showed that there is a critical period during which memories consolidate. In particular, they observed that learning a list of words (such as Ebbinghaus's) interfered with the learning of a previous list, and that the more the lists resembled each other, the more noticeable this interference was. This led them to infer that there is a critical

7. Eric R. Kandel shared the Nobel Prize in Physiology or Medicine in 2000 for his study of the processes through which memories consolidate. Avoiding a detailed discussion that would take us too far afield, let me just say that there are two different mechanisms for memory consolidation. The first involves protein synthesis, and the second is based on practice and repetition. In Argentina this is an active field of research, involving Héctor Maldonado's group at the Faculty of Exact and Natural Sciences, Carlos Baratti's laboratory at the Faculty of Pharmacy and Biochemistry, and Jorge Medina's lab at the Faculty of Medicine, all at the University of Buenos Aires.

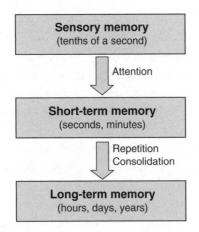

FIGURE 5.2
A simplified model of memory generation.

consolidation period during which the repetition of a certain short-term memory gives rise to the formation of a long-term memory. (See figure 5.2.)[8]

Nowadays we also know that sleep plays a dominating role in the repetition and consolidation process. This is why it is usually not advisable for a student to cram overnight before an exam, since lack of sleep will affect the consolidation of the material. The existence

8. We should note that this is a simplified model in which we have omitted the so-called working memory, the one we use to process information retained temporarily (for example during mental computation). We have also left out criticisms of this model based on certain pathological cases in which subjects without short-term memory have the capacity to form long-term memories starting directly from sensory memories.

of the consolidation period explains the not uncommon instances of temporary amnesia resulting from a brain contusion. For example, a boxer who has been knocked out usually fails to remember the details of the fight immediately preceding the knockout; epileptic patients are usually unable to remember the situation preceding seizures that include a loss of consciousness; most people who get seriously hurt in a traffic accident can barely recount what led to the accident. All these cases have the same underlying cause: the lack of consolidation of the memories preceeding the incident.[9] However, beyond this critical period during which memories set, their consolidation is a constant process. Repetition helps set memories even years after they are first acquired. For example, we tend to remember more vividly those events for which we have photographs that we occasionally look at. The photos make us revisit the events and reconsolidate the memories linked to them.

In the previous chapter we described H.M.'s case and how it proved that the hippocampus intervenes critically in the formation of long-term memories. H.M.'s contribution to our knowledge of memory does not end there, though. In a brilliant paper published in the early 1960s, Brenda Milner demonstrated the existence of a completely different type of memory. With a simple experiment, Milner showed that H.M. was capable of learning a complex motor task. She asked him to draw a line within the contour bounded by two concentric stars (the shaded region in figure 5.3), but guided only by a mirror reflection. The difficulty of this task is obviously

9. In fact, H.M. was unable to remember events from the years just preceding his surgery.

FIGURE 5.3

Experiment used by Milner to assess H.M.'s motor memory capacity. The task consists of following the contour bounded by the two stars using a mirror reflection as guide.

given by the visual feedback, which inverts the direction of motion —leftward motion looks rightward and vice versa. Surprisingly, not only was H.M. able to carry out the task without any problem, he also improved his performance with each passing day (as Miller ascertained by measuring the time it took him to complete the task and the number of mistakes he made). This demonstrated the existence of a memory for motor tasks, unaffected by H.M.'s surgery, that we currently call procedural memory: the one we use when we ride a bicycle, tie our shoelaces, brush our teeth, and so on.

H.M.'s capacity to learn this motor task—following with a pencil a contour bounded by two stars—contrasts starkly with his inability to remember new people or facts. Brenda Milner wrote that H.M. was surprised at his ability to take the test, given that he could not remember his previous learning sessions. He was not aware of his new skill. Milner concluded that there is a procedural or implicit memory that does not depend on the hippocampus, in contrast with the memory for facts and concepts, known as declarative or explicit memory, that does depend on the hippocampus and was severely compromised in H.M.'s case. Since Milner's time, declarative memory has itself been subdivided into semantic (the one for people, concepts, and places) and episodic (the one for events, which for example allows us to recall going to the movies with a friend).

We have seen that memory, far from being a monolithic and indivisible entity, is actually full of nuances. To simplify, I refrain from discussing several further subtleties, like working memory and other forms of implicit memory.[10] Let us then finish this chapter by going over the most salient ideas about the different types of memory.

The processing of information from the external world starts with sensory memories, which in the case of visual information last for only a fraction of a second. Starting with this representation, those things to which we devote our attention become short-term memories. These memories, which last for a few seconds, are the ones we use to dial a telephone number, construct a coherent sentence, or answer a question (since as we respond we have to keep remembering what the question is). Short-term memories are processed mostly by the frontal cortex, an area that was left untouched in H.M.'s case; this is why he had no problem carrying on a

10. The interested reader will find more details in:
• Larry R. Squire and Eric R. Kandel, *Memory: From Mind to Molecules* (Greenwood Village, CO: Roberts, 2008);
• Alan D. Baddeley, Michael W. Eysenck, and Michael C. Anderson, *Memory* (Hove, East Sussex, UK, and New York: Psychology Press, 2009).

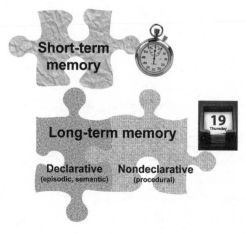

FIGURE 5.4
Types of memory.

coherent conversation with his interlocutors. Some short-term memories—though not all, or we would end up like Funes the memorious—can endure if they become long-term memories. Among these long-term memories there are nondeclarative ones (procedural memories, as well as others that we have not discussed for the sake of simplicity) and declarative ones (episodic and semantic). Procedural memories involve the basal ganglia and the cerebellum, among other brain areas. These regions were intact in H.M.'s case, and for that reason he had no problem learning and remembering new motor tasks. Declarative memories, on the other hand, are distributed all through the cerebral cortex—also intact in H.M.'s brain—but are formed from short-term memories with the help of the hippocampus. This explains why H.M. could remember past events but was incapable of forming new declarative memories.

WHERE DO MEMORIES RESIDE?

Borges masterfully used enumeration to describe the essential traits of a character, thus bringing the reader immediately to the thick of a plot. He begins "Funes the Memorious" thus:

Lo recuerdo (yo no tengo derecho a pronunciar ese verbo sagrado, sólo un hombre en la tierra tuvo derecho y ese hombre ha muerto) con una oscura pasionaria en la mano, viéndola como nadie la ha visto, aunque la mirara desde el crepúsculo del día hasta el de la noche, toda una vida entera. Lo recuerdo, la cara taciturna y aindiada y singularmente remota, detrás del cigarrillo. Recuerdo (creo) sus manos afiladas de trenzador. Recuerdo cerca de esas manos un mate, con las armas de la Banda Oriental; recuerdo en la ventana de la casa una estera amarilla, con un vago paisaje lacustre.

Recuerdo claramente su voz; la voz pausada, resentida y nasal del orillero antiguo, sin los silbidos italianos de ahora.[1]

[I remember him (I have no right to pronounce that sacred verb, only one man on Earth had the right and he is dead) with a dark passionflower in his hand, watching it like no one has watched it, even if he were to watch it from dawn to dusk, his whole life. I remember him, his face taciturn and native-looking and singularly aloof behind his cigarette. I remember (I believe) his sharp leather braider's hands. I remember a *mate* near those hands, with the cup showing the Uruguayan coat of arms; I remember a yellow mat with a vague picture of a lake in one of the windows of his house. I clearly remember his voice, slow, resentful, and nasal, a voice like they used to have along the riverbank, with none of the Italian hisses one hears nowadays.]

And this is how he starts "The Maker":

Nunca se había demorado en los goces de la memoria. Las impresiones resbalaban por él, momentáneas y vívidas; el bermellón de un alfarero, la bóveda cargada de estrellas que también eran dioses, la luna, de la que había caído un león, la lisura del mármol bajo las lentas yemas sensibles, el calor de la carne de jabalí, que le gustaba desgarrar con dentelladas blancas y bruscas, una palabra fenicia, la sombra negra que una lanza proyecta sobre la arena amarilla, la cercanía del mar o de las mujeres, el pesado vino cuya aspereza mitigaba la miel, podían abarcar por entero el ámbito de su alma.[2]

[He had never taken much time to enjoy the pleasures of memory. Impressions would just slide off him, ephemeral and vivid: the vermilion of a potter, a sky laden with stars that were also gods, the moon, from which a

---

1. Jorge Luis Borges, "Funes el memorioso," in *Obras completas* (Buenos Aires: Emecé, 2007), vol. 1, pp. 583–590.

2. Jorge Luis Borges, "El hacedor," in *Obras completas*, vol. 2, pp. 191–192.

lion had fallen, the smoothness of marble to slow and sensitive fingertips, the warmth of boar meat, which he liked to tear up with white, brisk bites, a Phoenician word, the black shadow that a lance projects on the yellow sand, the proximity of the sea or of women, the heavy wine whose roughness cut the sweetness of honey, could encompass the entirety of his soul.]

These compilations of memories take only a few lines to define two completely different characters: Funes, the peasant from Fray Bentos, Uruguay, and Homer, the poet from ancient Greece. We are our memories, say the folk. We are in fact much more, a scientist would argue, highlighting the value of instinct, imagination, and feelings. Beyond those technicalities, though, no one can deny the importance of memories in forging our ego and the awareness of our own self.[3] Where are these memories kept? Is there a specific area of the brain that stores what makes me who I am and distinguishes me from Funes and Homer?

The question of whether there is a place in the brain that is the seat of memory—or, more generally, whether the different brain functions are localized—has fascinated generations of thinkers over the centuries. It has its origins with the ancient Greek philosophers and their quest for the psyche, the human soul (stripped here of any religious connotation). Aristotle believed that the psyche originated in the heart. He reached this conclusion by analyzing the

---

3. In "La memoria de Shakespeare" ("Shakespeare's Memory"), Borges describes the misadventures of Professor Hermann Soergel, who is offered Shakespeare's memory. After his initial fascination, Soergel ends up rejecting this memory when he notices that it is melding with his own and that he is starting to lose his personal identity. Jorge Luis Borges, "La memoria de Shakespeare," in *Obras completas*, vol. 3, pp. 473–481.

motor system—how we move a finger, a forearm, a leg. The choice to study the motor system was not arbitrary but pragmatic, since it is easy to experiment with different movements. Aristotle erroneously ascribed our muscular movements, those manifestations of our free will and our psyche, to our blood vessels and the flow of blood that originates in the heart. Aristotle believed in the existence of a divine, incorruptible element called the *pneuma*, the air that surrounds us, the vital ether that gives rise to our psyche and leaves us when we cease to breathe it. According to Aristotle, this *pneuma* enters us through our lungs and then goes to our heart, where it becomes "vital *pneuma*" that travels through our blood vessels to reach our muscles. For Aristotle, the brain—a cold, inert organ

FIGURE 6.1
A bust of Aristotle (384–322 B.C.) and a portrait of Galen (Claudius Galenus, 130–200 A.D.).

when compared to the warm and active heart—played the role of cooling the blood, somewhat like a car's radiator.

Aristotle was eventually refuted by Galen (the extraordinary surgeon who gained part of his knowledge by performing surgeries on Roman gladiators), who discovered the nerves and their origin in the spinal cord and the brain. Galen refined Aristotle's model by deducing that the "vital *pneuma*" was taken to the brain by blood vessels and there turned into "psychic *pneuma*," which would become muscular movement by traveling through the nerves originating in the brain. The dispute was resolved once and for all by Galen's observations, though Alcmaeon and Hippocrates, who studied patients with cerebral injuries, and even Plato himself had proposed centuries earlier that the brain is the seat of the psyche.

Nowadays to think that the psyche can originate in the heart is as anachronistic as believing that the Earth is flat. It would be unfair, however, to dismiss Aristotle's reasoning offhand. Aristotle —the master of all sages, as Dante calls him in the *Divine Comedy*— did not have Hippocrates' or Galen's knowledge of anatomy and based his conclusions on pure thought. It was his prodigious reason that led him to postulate the fundamental principle of *monism*, the unity of psyche and body (or mind and matter). Aristotle believed that things were composed of matter and form. A statue is not a statue if it lacks the marble of which it is made, but it is also not a statue if it lacks the form that it represents. In Aristotle's line of reasoning (unlike that of Plato, his teacher, and unlike the later Cartesian dualism that we shall discuss further on), it is both the body and the psyche that make the animal. He considered it an absurdity to ask if the body and the psyche are one and the same thing, since it would be like asking if the wax that makes up a candle is the same as its shape. To talk about the psyche of a creature, says Aristotle, is to talk about the creature itself.

From Aristotle's reasoning one gathers that a creature will continue to have a psyche as long as it is alive, which, given the tendency to equate psyche with soul, would imply the mortality of the latter, an idea the Catholic Church had difficulty accepting.[4] Though Aristotle himself never explicitly denied the existence of an immortal soul or a God, it still took some 1,500 years for Christianity (i.e., Western thought) to accept his philosophy thanks to the interpretation offered by Thomas Aquinas (1225–1274). This resistance of medieval scholastic philosophy to coming to terms with Aristotle was a consequence of the writings of the Muslim philosopher Averroës (1126–1198, the character in Borges's "Averroës's Search"),[5] who, starting from Aristotle's monism, denied the immortality of the soul. According to Averroës, at the moment of death the soul loses its individuality and becomes part of a universal soul, like a drop of water in the ocean. Thomas Aquinas, on the other hand, took up Aristotle's distinction between active intellect (the one, exclusive to humans, that allows reasoning) and receptive intellect (the one we share with animals and that allows perception) and stated that the receptive intellect is the one that disappears upon death, while the active intellect, the soul itself that only human beings possess, is indeed immortal.

4. The opinions and positions of the Catholic Church are no triviality, given that its precepts dominated Western philosophy for centuries.

5. In Jorge Luis Borges, "La busca de Averroes," in *Obras completas*, vol. 1, pp. 743–756.

Breaking with the Aristotelian tradition that propounded the unity of body and soul, René Descartes took up again the idea, already set forth by Plato, that there exists a dualism, the famous Cartesian dualism, between mind (the soul, the psyche, the spirit) and matter (the body, the brain itself). Descartes postulated that the brain (in humans as well as in animals) takes care of reflexive acts, while the soul is in charge of mental processes. According to this model, the interaction between mind and body—the thinking that

FIGURE 6.2

René Descartes (1596–1650), shown here in an oil painting by Frans Hals that hangs at the Louvre, and one of his diagrams illustrating the function of the pineal gland. Light penetrates through the eyes and into the pineal gland, at the center of the brain, where the mind resides. This is where material stimuli turn into thoughts, which in turn give rise to intelligent actions (or muscle responses) in human beings.

arises from sensory experience, for example—occurs in the pineal gland, a central, unique organ (since everything else in the brain comes in pairs, one for each hemisphere) that at the time was erroneously believed to exist only in humans. Cartesian dualism, the division between mind and brain, dominated Western philosophy for almost two centuries, perhaps because its greater affinity with Christian beliefs made it much easier to accept than Aristotelian monism. The underlying reason is simple: the body is a material entity that lies within the domain of science, while the soul, the spirit, belongs to religion.

Despite the fact that Cartesian dualism extended its influence well into the twentieth century, I believe there are but few people with a minimum of scientific knowledge who would still doubt that the mind and the brain are one and the same thing.[6] Again I remark that, following Thomas Aquinas, this does not imply the nonexistence of an immortal soul. Aristotle, despite being wrong about the location of the psyche, laid the foundation of our current monistic view of the brain and thought. He wrote in *On the Soul* (408b):

6. However, Diego Golombek, the distinguished Argentine scientist and popularizer, argues—correctly, I believe—that brain research in Argentina has been long dominated by psychoanalysis, which almost stoically believes in a form of Cartesian duality and erroneously distinguishes between mind and brain, between mental and bodily illness, when the truth is that mental processes—depression, addiction, schizophrenia, or epileptic seizures, to name a few—arise from neuronal activity in the brain. As Diego says, in Argentina there seems to be a need for "more James and less Lacan." Diego Golombek, *Cavernas y palacios: En busca de la conciencia en el cerebro* (Buenos Aires: Siglo veintiuno, 2008), p. 35.

Yet to say that it is *the soul* which is angry is as inexact as it would be to say that it is the soul that weaves webs or builds houses. It is doubtless better to avoid saying that the soul pities or learns or thinks, and rather to say that it is the man who does this with his soul.[7]

To highlight the synergy with current scientific views (which moreover incorporate the fact that the brain is composed of neurons), I quote Francis Crick, who shared the 1962 Nobel Prize for Physiology or Medicine with James Watson and Maurice Wilkins for discovering the double-helix structure of DNA. Crick devoted the final 20 to 30 years of his life to studying the problem of consciousness. In the first paragraph of the first chapter of his fascinating book *The Astonishing Hypothesis*, he makes it clear that

"You," your joys and your sorrows, your memories and your ambitions, your sense of personal identity and free will, are in fact no more than the behavior of a vast assembly of nerve cells and their associated molecules.[8]

Along the same lines, Patricia Churchland, a philosopher and neuroscientist at the University of California, San Diego, states that, in the same way that electricity *is not caused by* the motion of electrons but *is itself* the motion of electrons, and temperature *is not caused by* the kinetic energy of molecules but *is itself* the kinetic energy of molecules, consciousness (to put it simply, a modern

7. Aristotle, *De anima* (*On the Soul*), trans. J. A. Smith, in *The Basic Works of Aristotle*, ed. Richard McKeon (New York: Random House, 1941), p. 548.

8. Francis H. C. Crick, *The Astonishing Hypothesis: The Scientific Search for the Soul* (New York: Simon and Schuster, 1993), p. 3.

version of Aristotle's psyche) *is not caused* by the activity of neurons but *is itself* the activity of neurons.[9]

These philosophical discussions have led us somewhat away from our initial question: Where do memories reside? I have let myself into this detour, and with pleasure, because I imagine it more akin to Borges than a dry description of scientific evidence about the localization of memories. Borges loved literature and philosophy, and it was through the humanities that he immersed himself in problems that nowadays attract scientists. Let us then continue this game of avoiding the question (as a good politician would) and, before coming back to the problem of where memory is located, let us ask ourselves a more general question: does the brain—which, as we have seen, gives rise to the mind—behave like a single, indivisible entity, or do its different areas perform different tasks?

The idea that each distinct, discrete area of the brain may be in charge of a different task was originally proposed by Franz Josef Gall, a German physician and neuroanatomist who studied how the size and shape of the cranium reflect different brain functions, a discipline that came to be known as phrenology. Assuming that the abnormal development of a given brain area would manifest itself as a protrusion in the cranium, Gall examined the skulls of people with different talents, flaws, or personalities, seeking for common patterns that would reveal anomalies in particular areas. He began this at a young age when he first observed that his smartest

9. Patricia S. Churchland, "Can Neurobiology Teach Us Anything about Consciousness?," in Ned Block, Owen Flanagan, and Güven Güzeldere, eds., *The Nature of Consciousness: Philosophical Debates* (Cambridge, MA: MIT Press, 1997), pp. 127–140.

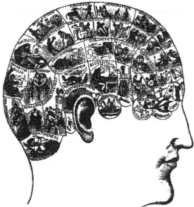

FIGURE 6.3

Franz Gall (1758–1828) examining a patient; a map showing the locations of different brain functions according to phrenology.

classmates tended to have bulging foreheads and eyes. Gall reasoned that this was due to an exceptional development in the underlying regions of the brain. After examining innumerable skulls, Gall postulated the existence of 27 different areas dedicated to functions such as memory, language, computation, sound processing, and even affection, hope, idealism, belligerence, religion, and faith.

As one would expect, the idea that mental functions originate in different parts of the brain did not please the Church and its embrace of Cartesian duality, which clearly separated the domains of science and religion and advocated a mind that was whole, indivisible, immortal, and separate from the body. In the same way that Aristotle was refuted by the Church for centuries, Gall was banned by Emperor Francis II from lecturing and practicing medicine at the

University of Vienna (where he had worked from 1781 to 1802). Gall continued teaching in different European cities, becoming a popular and influential figure, until he eventually settled down in Paris.

Gall's most vehement opponent was French neurologist Pierre Flourens (1794–1867), who earned his medical degree before he turned 20 and dedicated his career to proving Gall wrong (perhaps because he felt uncomfortable with the religious implications of Gall's ideas). Using dogs, rabbits, and birds as subjects, Flourens removed one by one each and every brain area that Gall had postulated to relate to different brain functions, and observed that their removal did not generate the specific changes in behavior attributed to them. In fact, Flourens observed that such lesions resulted in a global deficiency rather than in specific deficits, and that what really mattered was the extent of the lesion, not its location. These results led him to reaffirm that the brain behaves like an indivisible whole and that its functions are distributed all over its hemispheres.

Ironically, even though Flourens's experimental evidence was overwhelmingly more solid than Gall's—after all, a protrusion on the skull may be simply due to a bump—we now know that Gall's ideas were essentially correct. The problem with Flourens's experiments was that the lesions that he produced were insufficiently specific to localize different cognitive processes. In fact, the localization of brain functions is much more subtle than what Gall proposed (and which misled Flourens's experiments).

In Gall's time, phrenology boasted scholarly societies, elaborate "cephalographs" with which to measure crania, and highly influential professors dedicated to its study; nowadays the term "phrenology" is used dismissively to denote bad science. We should not, however, underestimate the influence of phrenology in the development of neuroscience, though more solid experimental support was needed before the idea of localized brain functions could be taken seriously.

In 1825, French physician Jean-Baptiste Bouillaud presented cases of patients who had lost their speech after suffering injuries in their temporal lobes. Bouillaud's work was largely ignored by the scientific community, but four decades later it was the spark that detonated one of the most important discoveries in the history of neuroscience. At the 1861 conference of the Société d'Anthropologie de Paris, Ernest Aubertin, Bouillaud's student and relative, presented one of Bouillaud's cases along with several similar ones that he himself had studied. Aubertin's presentation caught the attention of one of the Société's founders, Paul Broca (1824–1880), a brilliant French surgeon who specialized in the anatomy of the brain. That same year, Broca described the case of Leborgne, a 51-year-old Parisian cobbler who had lost the ability to speak after sustaining a brain lesion at the age of 30. The staff at the hospital had nicknamed him "Monsieur Tan" because "tan" was the only syllable he could mutter. Tan had no problem understanding language but could not articulate a fluent discourse—either orally or in writing—even though he could readily move his tongue, lips, mouth, vocal cords, etc. In fact, he could whistle or hum a melody without difficulty. Monsieur Tan died a week after having been examined by Broca, and a brain autopsy revealed a marked lesion on his left frontal lobe (see figure 6.4) that coincided with the previous descriptions by Bouillaud and Aubertin. Broca corroborated this finding with eight more patients, thus locating what we know today as Broca's area. This, the first irrefutable experimental evidence of the localization of specific brain functions in specific areas of the

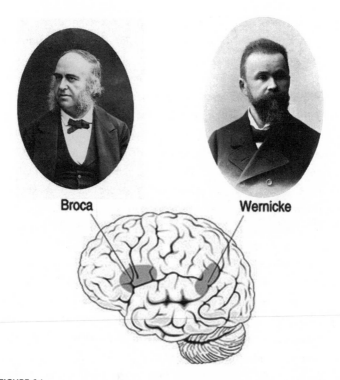

FIGURE 6.4

Paul Broca (left), Carl Wernicke (right), and the locations on the left hemisphere of the areas named after them. Broca's area, on the frontal lobe, is related to speech execution, while Wernicke's, located further back, is related to speech comprehension.

brain, led Broca a few years later to pronounce his famous statement "nous parlons avec l'hémisphère gauche [we speak with the left hemisphere]." Broca is now considered the founder of neuropsychology—the study of mental afflictions due to brain injuries.

Broca's findings were complemented and augmented by German neurologist Carl Wernicke (1848–1904), who in 1879 described the opposite phenomenon: patients who could speak (even if only incoherent phrases) but were unable to understand spoken or written language. When he performed autopsies on these patients, Wernicke observed that their temporal lobes consistently showed similar types of lesions. Tying his findings to Broca's, Wernicke proposed that language is processed by two connected areas (see figure 6.4): what we know today as Wernicke's area, in charge of language understanding, and Broca's area, in charge of its execution. Lesions to Wernicke's or Broca's areas lead to different pathologies, known respectively as "sensory aphasia" and "motor aphasia." Patients with motor (or Broca) aphasia can only utter isolated words, while those with sensory (or Wernicke) aphasia have no problem speaking but generally say only nonsense. In some cases the injury affects both areas and produces what is known as mixed aphasia.

Gustav Fritsch and Eduard Hitzig, two nineteenth-century German physiologists, added more evidence about the localization of brain functions when they demonstrated, by experimenting on dogs, the existence of an area devoted to the generation of movements (known today as the motor cortex). Fritsch and Hitzig saw that stimulating a relatively central area of the brain with small electric currents could produce different twitches on the face and limbs—on the opposite side of the hemisphere being stimulated—and that the locations of the twitches varied with the stimulated area. Moreover, they identified subareas within the motor cortex that were involved in moving different muscles.

These results were confirmed by the observations of John Hughlings Jackson (1835–1911) on epileptic patients. Jackson noticed that in many cases epileptic seizures were limited to one side of the body, and subsequent autopsies revealed anomalies on the opposite side of the brain, which led him to believe that the epileptic activity was due to pathologic activity in the opposite hemisphere. Moreover, Jackson observed that the twitches that occurred during epileptic seizures had time-dependent patterns: a given seizure could start with twitches on one side of the face and continue with hand movements that would then propagate to the forearm and so on. This led him to postulate a topographic arrangement of the motor cortex: the temporary activation of different groups of muscles was due to the seizure's spread to the contiguous brain areas that controlled their motion. Sir David Ferrier, a Scottish scientist, extended Jackson's clinical observations when he demonstrated, using specific stimuli and lesions, the existence of a topographic arrangement of the motor cortex in monkeys. Later, Sir Charles Sherrington (the mentor of Wilder Penfield, whom we shall meet later) used electric stimulation in monkeys to localize the motor cortex more precisely.

So far we have been discussing the localization of brain functions. We initially described the ideas of ancient Greeks, who first believed that thought originated in the heart and then that it originated in the brain; we continued with Descartes and his distinction between mind and body, and we followed the debate about phrenology until reaching Broca and those who followed him. This may seem an outdated discussion, but the localization of brain functions continues to be enormously important for neuroscience. In fact, many researchers dedicate their whole careers to finding the areas of the brain involved in different functions, using techniques like functional magnetic resonance imaging. Let us now go back to our initial question: Is there a place in the brain that houses the memory of Funes, with a dark passionflower in his hand, his native-looking face and his sharp leather braider's hands? Is there another area holding the vermilion of a potter, the sky laden with stars, and the warmth of boar meat?

The debate about the localization of brain functions, which began with Gall and Flourens in the mid-nineteenth century, sooner or later had to move on to the study of memory. However, in spite of a wealth of evidence showing that language and motor processes were localized, in the second half of the twentieth century, almost a century after Broca's discoveries, it was still believed that memory was distributed all over the brain. This "holistic"[10] view of memory was defended by Karl Lashley (1890–1958), a very influential professor of psychology at Harvard University, who reached this conclusion after failing to find an area in the rat brain where memory would preferentially reside. Based on the findings of Broca, Wernicke, and others, Lashley trained rats to navigate a simple maze and studied how their ability to go through it—their spatial memory—was affected as he performed lesions in different brain areas. Just like Flourens a century before him, Lashley found that the rats' memory loss depended on the size of the lesion rather

10. This is analogous to a hologram, in which each point contains information about the whole image.

than on its location. But, also just like Flourens, he made two fundamental mistakes: first, he injured only the cerebral cortex, leaving intact internal structures—like the hippocampus—that we know today are essential for the formation of memories; second, the technique he employed was inadequate to assess memory capacity because it was insufficiently specific and comprised too many sensory and motor processes. In short, if a given lesion affects the area that controls one of the senses (e.g., touch), the rat can still use its other senses (like sight and smell) to move through a maze.

As we saw in previous chapters, Lashley's idea of a holistic memory was refuted by Milner's findings with Patient H.M. However, the holistic theory was so established in the scientific community that it was only after many years, following independent verification with other patients and experiments with monkeys replicating H.M.'s lesion, that Milner's ideas started to gain acceptance. Milner proved that memory is not an all-encompassing distributed system by showing, first, that there are different types of memory, and second, that a localized region of the brain, the hippocampus, plays a crucial role in the generation of declarative memories—those dealing with concepts and events. The hippocampus is not where (factual) memories are *stored*, since H.M. could still remember events that preceded the surgery in which both his hippocampi were removed. Earlier we made the comparison to a clerk whose job consists of storing new memories in the cerebral cortex. If the hippocampus is missing, then new memories fail to be stored and inexorably fade away.

We have covered much ground and, although we have not yet answered our initial question, we can now address it more precisely. Wilder Penfield, who was Brenda Milner's mentor and studied with her two cases similar to H.M.'s, made spectacular breakthroughs in this regard. We have already said that some epileptic patients who do not respond to medication can be subjected to surgery to remove the area that causes the seizures. The key to the success of these surgeries is twofold: first, it is essential to locate the epileptic focus as precisely as possible; second, one must make sure that the focus is not in a vital area (for example, if part of the motor cortex is removed the patient may end up paralyzed). The locations of these vital areas vary from patient to patient, and for that reason Penfield developed a procedure whereby the patient's brain would be electrically simulated during surgery, with the patient awake and the brain fully exposed. This requires only local surgery, since the brain itself has no pain receptors; the main difficulty lies, of course, in training the patients to stay calm and not panic despite their clinical

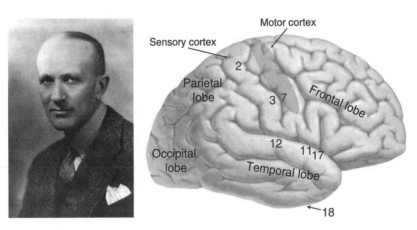

FIGURE 6.5
Wilder Penfield and the areas that he stimulated on Patient M.M.'s right hemisphere.

situation. During these procedures, Penfield stimulated around Broca's and Wernicke's areas to make sure he would not injure them with his scalpel (and cause the patient to lose speech). To determine the extents of these areas, Penfield would ask the patient to name different objects (hence the need for the patient to be awake) as he applied the electrical stimulus. The stimulation of Broca's and Wernicke's areas compromised the patient's language skills, thus enabling Penfield to demarcate them exactly.

Beyond its clinical importance, his technique of applying electric stimuli on awake patients made it possible to determine the roles played by different areas of the brain. Following the work of Fritsch, Hitzig, and Ferrier which we described above, and after studying more than a thousand patients with this method, Penfield constructed a map of the sensory and motor cortices of the human brain. When he stimulated these areas, Penfield would cause tickling and induce movements of different body parts like the fingers, the tongue, or the lips. When he stimulated the zones devoted to the processing of visual information, the patients would report experiencing visual hallucinations, such as colored stars. And when he stimulated the temporal lobe—and only there—Penfield could make his patients recall memories. The results are so remarkable that they are worth describing in detail.

Penfield confessed having felt both disbelief and wonder the first time (in 1933) he saw patients reliving their past, as in a movie flashback, when he stimulated their temporal lobes.[11] For example,

---

11. Wilder Penfield, *The Mystery of the Mind: A Critical Study of Consciousness and the Human Brain* (Princeton: Princeton University Press, 1975), pp. 21–27.

a patient said she could remember being in her kitchen, listening to her son play outside. She could recall the sounds of cars passing by in the neighborhood and was aware that they could be dangerous for her son. Another patient heard a melody whenever a particular area of his brain was stimulated. Penfield repeated the stimulus 30 times and each time the patient repeated the same melody from the same point on. Then there was Patient M.M. Penfield stimulated different points of her right brain hemisphere (see figure 6.5) with the following results: when point 2 was stimulated, the patient reported feeling a tickle in her left thumb; at point 3 she felt a tickle in the right-hand side of her tongue; point 7 caused a twitch of the tongue. Clearly, points 2 and 3 were on the sensory cortex, while point 7 was on the motor cortex. The most interesting part, however, resulted from stimulating M.M.'s temporal lobe (points 11, 12, 17, and 18):

- Point 11 [The patient reports]: "I heard something, I do not know what it was."
- Point 11 [Penfield repeated the stimulation without warning]: "Yes, I think I heard a mother calling her little boy somewhere. It seemed to be something that happened years ago. . . . It was somebody in the neighborhood where I live."
- Point 12: "Yes. I heard voices down along the river somewhere—a man's voice and a woman's voice calling . . . I think I saw the river."
- Point 17: "Oh! I had the same very, very familiar memory, in an office somewhere. I could see the desks. I was there and someone was calling to me, a man leaning on a desk with a pencil in his hand."
- [Penfield warns her that he will stimulate her brain, but does not]: "Nothing."

- Point 18 [At the bottom of the temporal lobe; Penfield applies a stimulus without warning]: "I had a little memory—a scene in a play—they were talking and I could see it—I was just seeing it in my memory."

Nowadays nobody doubts the accuracy of Penfield's mapping of the sensory and motor cortices (also called the sensory and motor "homunculi"). However, the remarkable results that he obtained when he stimulated the temporal lobe are still the subject of debate.[12] Penfield argued that memories reside in the temporal lobe and that its stimulation evokes them and triggers the flow of consciousness, in the same way that the taste of a madeleine triggered Marcel Proust's memories in *Remembrance of Things Past*. However, such evocation worked only in 8 percent of Penfield's cases. Moreover, in 40 percent of the cases where he sparked memories by stimulating the temporal lobe, these were identical to the hallucinations experienced by the patients during their epileptic seizures; this led Penfield's critics to argue that, rather than triggering the flow of consciousness, he was simply eliciting seizures. Besides, in

12. Speaking of remarkable results, Itzhak Fried (a neurosurgeon with whom I have worked for years studying neural recordings from epileptic patients, and a great admirer of Penfield) showed that stimulating the supplementary motor cortex of a patient would consistently make her laugh. When asked (during the surgery) what the laughter was about, the patient said to the surgeons: "you guys are just so funny . . . standing around." More details can be found in Itzhak Fried, Charles L. Wilson, Katherine A. MacDonald, and Eric J. Behnke, "Electric Current Stimulates Laughter," *Nature* 391 (1998): 650.

many cases these reminiscences included fantastic, dreamlike elements that did not correspond to real memories.

On the other hand, one can still argue that, even if stimulation worked in only a few cases, and even though in some of these cases the evocation merely reproduced the pattern of an epileptic seizure, the very fact that there have been cases like those described by Penfield is extraordinary. Does this mean that memories reside in the temporal lobe? Probably yes, especially for visual and auditory memories, because the temporal lobe processes those types of information. Similarly, one would expect that memories related to touch—say, the texture of wool or leather—would be located in the sensory cortex (where this type of information is processed), and memories of smells in the olfactory cortex. This reasoning is perhaps speculative at this point, though it seems to make sense. When we recognize someone's face we are actually comparing the visual information with our memories of that person's face. One would reasonably expect that the memory of the face would be stored in the same place where we process information leading to its recognition. More elaborate memories—think of a flower, with its smell, texture, shape, and color—would be distributed over different brain areas. In the end, memories indeed seem to be distributed, though in a very different way than Flourens or Lashley envisioned. Not all areas work in the same way when it comes to storing memories; rather, each one captures a different piece of the whole, be it Funes's taciturn and aloof face, his sharp leather braider's hands, or his voice, resentful, slow, and nasal, like they used to have along the riverbank.

PRODIGIOUS MINDS

At the Cincinnati airport, Raymond Babbitt (Dustin Hoffman), an autistic man with savant syndrome, refuses to take a plane to get to Los Angeles. His brother Charlie (Tom Cruise), who barely knows him, tries to change his mind by telling him that flying is the safest way to travel. Raymond does not budge:

Raymond Flying's very dangerous. In 1987, there were 30 airline accidents. Two hundred and eleven were fatalities and 231 were definitely passengers. . . .

Charlie Is it this airline?

Raymond Yeah.

Charlie Okay, fine. There's an American plane.

Raymond American flight 625 . . . crashed April 27, 1976.

Charlie [checking the flight schedule] We don't have to take American. There's a lot of flights.

Raymond Yeah. Pick another airline.

Chapter 7

Charlie   Continental. We'll take a Continental flight.
Raymond   Continental crashed November 15, 1987. Flight 1713. Twenty-eight casualties. . . .
Charlie   Now, there's a Delta.
Raymond   Yeah.
Charlie   It leaves at midnight. How's Delta?
Raymond   Delta crashed August 2, 1985. Lockheed L-1011. Dallas-Fort Worth. Terrible wind shear. One hundred and thirty-five passengers.

Sensing Charlie's despair, Raymond proposes flying Qantas, the only airline that "never crashed." Charlie explains that Qantas has no direct flights from Cincinnati to Los Angeles and that to take Qantas they would first have to get to Melbourne, Australia. "Canberra's the capital," notes Raymond. "Sixteen point two million population. Very lovely beaches." Charlie, overwhelmed, decides to travel by car, and in that six-day road trip across the United States he will get to know his autistic brother in Barry Levinson's famous movie *Rain Man*.

The scene we just summarized gives a good idea of how people with savant syndrome behave. Darold Treffert, a consultant for the film and one of the foremost authorities on the topic, defines it as a condition whereby people with a serious mental deficiency (such as autism) have spectacular "islands of genius" that contrast markedly with their condition. In another scene from the film, Charlie asks Raymond to multiply 312 by 123 and he immediately gets the right answer: 38,376. He then asks how much is 4,343 multiplied by 1,234, and again, amazingly, Raymond gives the correct response: 5,359,262. The square root of 2130? Raymond is right once more: 46.15192304. But then Charlie asks, "If you had a dollar, and you spent 50 cents, how much money would you have left?" "About 70 cents." "Do you know how much a candy bar costs?" "About $100." "A hundred dollars, huh? You know how much one of those new compact cars costs?" "About $100." Like many savants, Raymond Babbitt could remember an incredible amount of facts and carry out difficult computations, but he was incapable of realizing that if he had a dollar and spent 50 cents he would have another 50 cents left.

In 1887, British psychiatrist John Langdon Down gave a series of talks at the London Medical Society. In one of them he described the developmental condition that now carries his name. During these talks he also presented 10 cases similar to Raymond Babbitt's, whom he named "idiot savants" (erudite idiots),[1] and called attention to the striking contrast between their disabilities in some areas and their genius in others. As an example he described the case of a boy who learned by heart Gibbon's *Decline and Fall of the Roman Empire*. Down called this talent "verbal adhesion": the ability to repeat something from memory without understanding its content—memory without consciousness. As with Shereshevskii or Ireneo Funes, the distinctive feature of savants is a prodigious memory accompanied by a severely limited capacity for general or abstract reasoning.

Rain Man is in fact based on the true story of a remarkable savant, Kim Peek, who during a chance encounter deeply impressed a Hollywood screenwriter (who went on to script the film and win an Oscar for it). Peek had an astonishing memory, perhaps the most

---

1. We note that in Down's time the word "idiot" did not possess the pejorative sense that it does now.

extraordinary on record. He would amaze people by asking them their birth date and immediately and without hesitation telling them the day of the week they were born, the day of the week their birthday would fall that year, and the day of the week they would turn 65 and thus be eligible for retirement. He knew the zip codes and area codes of thousands of U.S. towns, as well as the names of their local television stations and nearby highways. He had an unlimited capacity to recall historical facts from the last two millennia. He could tell when Mozart was born, when he died, and when he wrote each of his symphonies; he could name all the British monarchs in the correct order or tell the date of any baseball game. He could answer any question on American and world history, the lives of world leaders, geography (he knew the names of streets and routes in the United States and Canada), films, actors and actresses, sports, literature, music (he could identify every piece of music he had ever heard and tell its date of composition, name its composer, and give the composer's dates of birth and death), stories from the Bible, and so on. Students, children, and academics challenged him in public on innumerable occasions, and he always gave the right answers. He read constantly, dozens of books a day, and in less than ten seconds could simultaneously read two pages, one with each eye.[2] He once proved that he could read eight pages in 53 seconds and could recall more than 98 percent of what he had read. It was estimated that he knew the content of some 12,000 books.

2. It is believed that Peek's capacity to read a different page with each eye was due to the absence of the corpus callosum—the connection between the two hemispheres of the brain—which was noticed when his brain was imaged.

We mentioned earlier that Pliny the Elder described in his *Naturalis historia* several men with extraordinary memory: King Cyrus of Persia, who knew the names of all his soldiers; Mithridates, who spoke the 22 tongues of his empire; Charmadas, who could recite a book as though he was reading it. Funes, in Borges's story, could not understand that someone would be remembered for something that seemed so ordinary to him. I dare to think that Kim Peek (who died in December 2009 at the age of 58) might have felt the same way, since his talent greatly exceeded those of these historical figures. It is difficult to conceive that someone could have such memory capacity, but, unlike for the characters described by Pliny, there are scores of videos and scientific papers that give testimony of his feats.[3]

Kim Peek was like a computer that could not filter information— a sort of walking Google. (In fact, many called him *Kim-puter*.) However, and just like Shereshevskii or Funes himself, he had a limited capacity for reasoning. Studies by psychologist Rita Jeremy of the University of California, San Diego, showed that Peek had trouble solving problems that involved new reasoning and could not be solved based only on previous memories. Similarly, Vilayanur Ramachandran (also at UCSD) showed that Peek was unable to understand metaphors since he processed words literally. For example, when Ramachandran asked him about the meaning of "all that glitters is not gold," Peek gave a nonsensical answer ("I'll

3. Many of these videos can be watched on YouTube. Peek's mnemonic capacity, and those of other savants, are also described in Darold Treffert's *Islands of Genius* (London: Jessica Kingsley, 2010). Treffert interacted with Peek for many years.

give you a pound of my flesh if you don't watch out"), and when asked to explain the meaning of "George Bush is not exactly a rocket scientist," Peek responded that "he's the president of the United States." At Ramachandran's insistence, Peek alleged that "[Bush] doesn't know about how to look his message in the stars," again nonsense.

It is not surprising, then, that Kim Peek would not read fiction or books that require imagination, reasoning, or any faculty beyond plain memory; he read only books with real facts, devoid of ambiguity or multiple interpretations. When Ramachandran asked him to memorize a list of words related to the concept "sweet" (cake, honey, candy, sugar, chocolate, etc.) but that did not contain the word "sweet" itself, Peek correctly answered that "sweet" was not in the list, unlike most people (about 98 percent), who tend to abstract the concept and (incorrectly) include the word as part of the list. Peek would store information literally, without processing it to form abstract concepts, or, as Langdon Down would say, by simple verbal adhesion.

This is reminiscent of Luria's experiment with Shereshevskii (described in chapter 3) in which S. was given a list of words that included the names of several liquids. Shereshevskii could repeat the list without error but could not name the liquids in the list. Unlike Shereshevskii, who had to make an effort to forget something, Kim Peek simply did not forget. In fact, he kept acquiring information at a frenetic pace until the day he died and never reached the limit of his storage capacity.

In his later years, Peek started to develop a talent for music. His father, however, had to refrain from taking him to concerts because he knew every score by heart and would be irked by the slightest mistake, however insignificant, and would reproach the conductor at the end of the concert. The most embarrassing situation happened when, midway through a Shakespeare play, Peek started yelling at the actors to stop performing. When one of the actors asked what the problem was, Peek said someone had skipped some words from a line; the actor then explained that he did not think anyone knew or cared, to which Peek retorted that Shakespeare would have.

I cannot finish my description of Kim Peek without mentioning his good heart and the extraordinary love of his father, who helped him to get on in the world and accompanied him on every daily visit to the library, every trip, every medical appointment, and every lecture. After the success of *Rain Man*, Peek became a celebrity and traveled millions of miles to talk at universities, schools, churches, and any other institution that would invite him. He would finish these talks by saying: "We are all different. You don't have to be handicapped to be different. Treat other people like you would like to be treated and the world will be a better place."

Like Peek, Daniel Tammet also became a sudden celebrity. On March 14, 2004, at the Museum of the History of Science, in Oxford, Tammet managed to recite from memory the first 22,514 digits of π in a little over five hours.[4] Daniel Tammet also suffers from savant

---

4. The irrational number π relates the radius r of a circle to its perimeter l by $l = 2\pi r$. The number cannot be expressed as a quotient of two integers; when written out as a decimal it starts with 3.14 followed by an infinite sequence of digits with no discernible pattern. The day of Tammet's presentation, 3/14, was purposely chosen.

FIGURE 7.1
Kim Peek (left; credit Darold A. Treffert, MD; Wisconsin Medical Society),
who inspired the film *Rain Man*, and Daniel Tammet (right), who recited
from memory the first 22,514 digits of π.

syndrome and possesses an amazing talent for memorizing
numbers and performing complex mathematical computations.
When asked by scientists to raise 37 to the 4th power (that is, multiply 37 by 37 by 37 by 37), he answered at once: 1,874,161. He also
computed the fraction 13/97 = 0.134020618556701 . . . with higher
precision than the calculator of the interrogating researcher, who
had to resort to a computer to verify the accuracy of Tammet's
answer.

Like Shereshevskii, Daniel Tammet possesses a high degree of
synesthesia: for him, each integer between 1 and 10,000 has its own
shape, texture, and color. In Tammet's case, however, this richness
of perception is combined with a form of autism called Asperger's

syndrome,[5] which makes him focus almost obsessively on particular topics like numbers and mental computations. Thanks to this combination of synesthesia and autism it took Tammet only three months to memorize more than 22,500 digits of π, which he imagined as a continuous landscape formed by the shapes of the numbers. Once he etched the landscape in his memory, all Tammet had to do was read out loud the numbers that composed it to an audience that watched in disbelief.

This goes some way toward explaining how Daniel Tammet could memorize π, but does not explain at all how he can compute. To understand this, we must remember that he grew up autistic. Numbers were always his passion, his obsession.[6] Whereas the rest

5. Asperger's syndrome is a moderate form of autism that manifests itself through difficulties in communication and social interaction, along with an obsessive interest in a very narrow field. Unlike people with other forms of autism, those with Asperger's have a normal IQ and no language or mental development problems.

6. In his fascinating book *Born on a Blue Day*, Tammet gives a first-person account of the difficulties brought about by his autism and his efforts to overcome them. The book is an enormously important document, given the extreme rarity of an autistic person being able to communicate with the outside world. Most books on autism have been written by professionals in the field who, no matter how expert, are still unable to get into their patients' heads in order to understand what it means and feels to be autistic. In his second book, *Embracing the Wide Sky*, Tammet gives a detailed account of his interactions with scientists and explains how he performs mental computations, learns a language, or memorizes π.
• Daniel Tammet, *Born on a Blue Day: Inside the Extraordinary Mind of an Autistic Savant* (London: Hodder and Stoughton, 2006);

of us might spend our days thinking about people, places, and happenings, Tammet spent his childhood almost completely alone, thinking about numbers. When he thinks of a number, say 23, he can immediately recall semantic relationships—such as that 529 is 23 squared, 989 is the largest multiple of 23 smaller than 1000, etc.[7] In the same way that we build relationships between beings and concepts—when we think of the word "jungle" we relate it to a lion, and we relate the lion to its roar, its mane, etc.—Daniel Tammet builds relationships between numbers. We are so used to the sheer number of operations that our brains execute that we do not get surprised by their complexity. In a world populated by savants like Tammet, we would be seen as disabled, since we would be unable to perform the operations that they carry out effortlessly; but we would also be seen as prodigies given our power to socialize and do other things that they can hardly achieve. Perhaps the big difference between savants and other people is not how they memorize, but what.

Daniel Tammet's incredible memory was studied by researchers at the University of Cambridge using conventional tests.[8] He was given sequences of eight digits that he had to repeat after an interval of several seconds. What is interesting about this experiment is that

---

- Daniel Tammet, *Embracing the Wide Sky: A Tour across the Horizons of the Mind* (London: Hodder and Stoughton, 2009).

7. Tammet, *Embracing the Wide Sky*, p. 190.

8. Daniel Bor, Jac Billington, and Simon Baron-Cohen, "Savant Memory for Digits in a Case of Synaesthesia and Asperger Syndrome Is Related to Hyperactivity in the Lateral Prefrontal Cortex," *Neurocase* 13 (2008): 311–319.

two types of sequences were used: one type featured ascending and descending patterns (8 6 4 2 3 5 7 9, for example), while in the other the numbers were chosen at random. Tammet recalled both types of sequences with the same accuracy and, surprisingly, he failed to notice that half of them had some structure, while 85 percent of ordinary subjects did notice the patterns and used them to their advantage. This brings to mind the experiment in which Shereshevskii used the "brute force" of his prodigious memory to memorize a table of numbers without realizing that they were consecutive. In the same paper, the Cambridge researchers showed that Tammet had difficulties with abstract thinking and a strong tendency to focus on details. To that end they used the Navon test, in which letters formed by other smaller letters are shown to the subject. Figure 7.2 shows two examples, a letter H made out of small A's and an A built of small H's. This tends to generate interference—the large letter makes it harder to identify the smaller one, resulting in longer reaction times—but Tammet responded quickly and accurately because he had difficulty perceiving the big letters.

```
A     A            H
A     A           HH
A     A          H H
AAAAAA          H   H
A     A         HHHH
A     A        H      H
A     A        H      H
```

FIGURE 7.2
Two examples of letters used in the Navon test.

Tammet's talent is not limited to numbers; he also has an uncanny ability with words. He speaks more than 10 languages and, in response to a challenge posed by a British television show,[9] proved that he could learn to speak Icelandic in only a week.[10] He also took up chess and made remarkable progress in a very short time. However, when he tried his luck at tournaments he had disappointing results because he could not concentrate on the games; any external noise or any movement by his opponent would distract him.[11] In other words, it was hard for him to filter irrelevant information and focus on the essential.

Like Shereshevskii, who would be distracted by minute changes in the voice and tone of whoever was talking to him, Tammet has difficulty grasping the content of a conversation. He can hardly read between the lines or understand metaphors or sentences that are ambiguous or have a complex structure (for example double negatives like "I would never do anything I don't want"). He cannot understand the general idea of what is being said to him, or even if he is expected to make a comment or answer a question. Like others with autism, he has problems sensing the mood of the person with whom he is interacting: whether this person is interested, annoyed, bored, or willing to change the topic. Surprisingly,

9. The show, called *Brainman* in obvious reference to *Rain Man*, is easily found on the Internet.

10. Icelandic happened to be one of Borges's favorite languages because of its kinship to Anglo-Saxon and because it remained pure for far longer than present-day German or English.

11. Tammet, *Born on a Blue Day*, pp. 114–116.

and in spite of his tremendous ability to remember numbers and perform complex calculations in his head, he is relatively poor at algebra because he cannot abstract and manipulate variables and equations as he does numbers.

The astounding memory of people with savant syndrome can also manifest itself in painting or music. Leslie Lemke, for example, blind from birth and suffering from severe cerebral palsy (to the extent that he cannot dress himself or eat with fork and knife), is an enormously talented pianist. When he sits at the keyboard his palsy disappears as if by magic and he can play countless pieces. His memory for music is such that at the age of 14 he flawlessly played Tchaikovsky's first piano concerto after having heard it only once. Like Lemke, Blind Tom, a nineteenth-century black musician, had a vocabulary of around 100 words but knew over 7,000 songs by heart. Another notable case is that of Stephen Wiltshire, whose extraordinary talent for drawing stems from a prodigious visual memory but who learned to talk only at age 9. On his very first visit to Tokyo, and after only a half-hour helicopter trip, Wiltshire produced from memory a city panorama with detailed renderings of hundreds of buildings, parks, and streets; the painting took a week to finish and occupies a piece of canvas 30 feet long. He has made similarly detailed and staggeringly accurate paintings of Rome, London, New York, Madrid, and other large cities, all from memory.[12] It is worth pointing out that, unlike

12. Words are incapable of describing the panoramas painted by Stephen Wiltshire. They are exhibited in his London gallery and can also be admired at his website, www.stephenwiltshire.co.uk.

other artists, who start with a general outline and then add details, Wiltshire draws each building in turn, window by window, feature by feature. In other words, he piles up detail upon detail to produce the whole.

The neurologist Oliver Sacks narrates his interaction with Wiltshire in his book *An Anthropologist on Mars*.[13] Impressed by Wiltshire's drawings, Sacks invited him to his house in the outskirts of New York City and after a brief walk outside asked him to draw a picture of the property. Wiltshire accurately reproduced a myriad of details, but what most surprised Sacks was that he did not start by sketching the outline of the house but instead started on one end of the sheet and proceeded through at a steady pace drawing one detail after another, as though he were reproducing a picture already drawn in his mind. Sacks also noticed that Wiltshire's visual memory did not require coherent figures or conventional structures, because he could as easily draw from memory scenes resulting from demolitions or earthquakes. In a later meeting Sacks gave Wiltshire a jigsaw puzzle, which he had no problem solving. Sacks then gave him a different one, with the pieces face down, and Wiltshire solved it just as effortlessly, guided only by the shapes of the pieces.

Again we see that savants have an amazing power to perceive and memorize details but cannot turn them into general, abstract

13. Sacks devotes some 40 pages of his book to a description of Wiltshire, with whom he met many times and shared many trips. Despite being close to him, Sacks describes Stephen's inability, typical among autistic people, to show affection or emotions. Oliver Sacks, *An Anthropologist on Mars: Seven Paradoxical Tales* (New York: Knopf, 1995).

concepts. Thus the verbal (and, we could add, visual) adhesion mentioned by Langdon Down: a limitless capacity for literal recollection but with poor understanding, a difficulty in extracting the main point and ignoring the rest, a limited ability for relating concepts and forming abstract constructs. A mathematical savant like Daniel Tammet is skilled at calculation but cannot understand—let alone prove—a theorem as a mathematician would. (Curiously, it is not hard to find instances of the opposite—brilliant mathematicians who struggle with simple calculations.)[14] On the other hand, savants' skills and memories tend to be deep but narrowly focused, usually circumscribed to areas that most people do not care about (which is perhaps what makes them more prominent). After all, why would anyone want to memorize thousands of digits of π,

14. When I was in college I once attended a talk by Enzo Gentile, one of the foremost Argentine mathematicians and the author of an algebra textbook that anybody going through the Faculty of Exact Sciences at the University of Buenos Aires had to study at some point. The talk was about one of the most famous and challenging topics in mathematics, Fermat's last theorem, whose proof—finally obtained in the mid-1990s after centuries of failed efforts—Fermat had omitted because he could not fit it in the margin where he was writing. Gentile discoursed about the theorem as naturally as an Englishman would discuss the weather over a cup of tea, until at some point he wondered how old Fermat was when he died. He remembered Fermat's years of birth and death, so all he had to do was subtract. He did it and continued his talk . . . but then he wasn't sure. He went back and tried again, hesitated, wrote the numbers on the blackboard, attempted the subtraction as if he were a pupil in elementary school, hesitated again, and then gave up and continued with the (till then impossible) proof of the theorem. . . .

know what day of the week June 7, 1942, fell on, or be able to name all the British monarchs in order?

Beyond these general aspects, nobody knows with certainty what kind of brain alteration gives rise to savant behavior. Some have posited that a brain hyperconnectivity grants these people explicit access to memories and processes that ordinary people store in their subconscious. Others believe that it may be due to a dominance of the right brain hemisphere, which controls artistic skills, spatial processing, mechanical activities, and concrete processing, compared to the left hemisphere, which dominates most people's brains and controls language processing, speech, and symbolic and abstract reasoning. This conjecture is supported by observed pathologies in the left hemispheres of some savants. For example, Orlando Serrell, a savant from Virginia, can remember the day of the week, the weather, and everything he did on every single day since he was hit very hard by a baseball on the left parietal lobe at age 10. What is interesting about this case is that savant behavior does not necessarily arise from genetic alterations but can also appear spontaneously in normal people, whether from a blow to the head, epilepsy, or the presence of a tumor.

THE DELICATE BALANCE BETWEEN REMEMBERING AND
FORGETTING

I remember—I must have been five or six—running on the beach toward the ocean and turning around at the last second to avoid the water. I remember having spent some time, which now seems endless, playing the same game over and over again. This is perhaps one of my first childhood memories. I would in principle like to extend its recollection and rescue from the labyrinths of my memory even a tiny further detail, however small, from that day at the beach. But there must be a reason why nature makes us forget most of our memories. Perhaps so as not to overcrowd our minds and have us be like Funes, or maybe just to make us feel a bit melancholy, as when we see an old and worn black-and-white photo or revisit our past when we listen to a particular song.

In *Until the End of the World*, a fascinating film by Wim Wenders, Sam Farber (William Hurt) crisscrosses the planet with a special camera that records not only images but also the brain reactions of whoever shoots them. His father, a scientist holed up in a remote laboratory in the Australian outback, then tries to transmit these signals directly to the brain of his blind wife. Once he manages this feat—a major achievement in itself—he attempts the inverse process: instead of embedding the camera's images in the brain, he tries to interpret the brain's signals so as to capture and record the images it produces naturally. What is interesting here, at least to me, is not the science fiction—though, as we shall see, not much fiction anymore—but the description of the obsession and the suffering that people feel when they see images from their dreams and, within them, lost memories from their past. As Borges once said in his interview with Antonio Carrizo, one should try not to lean too much on one's memories because they can become an illness or an addiction.

In fact, from Borges's description of Funes, from the experience with savants, or from Shereshevskii's case, it becomes clear that it is not healthy to remember more than necessary. As Borges puts it in "A Reader" (*In Praise of Darkness*, 1969):

Mis noches están llenas de Virgilio;
haber sabido y haber olvidado el latín
es una posesión, porque el olvido
es una de las formas de la memoria, su vago sótano,
la otra cara secreta de la moneda.[1]

[My nights are full of Virgil;
to have known and forgotten Latin

1. Jorge Luis Borges, "Un lector," in *Obras completas* (Buenos Aires: Emecé, 2007), vol. 2, pp. 450–451.

is a possession, because forgetting
is a form of memory, its vague basement,
the other, secret, side of the coin.]

In the first chapter we quoted William James, who in his *Principles of Psychology* wrote that if we remembered everything we would be as disadvantaged as if we forgot it all, and that, paradoxically, we must forget to be able to remember. Thus there must be a balance between remembering and forgetting: we do not want to end up like Funes, who remembered absolutely everything, or like Patient H.M., who could not retain anything new. Other authors besides James, Borges, and Luria have written about the importance of forgetting, among them Iván Izquierdo, author of *The Art of Forgetting* (*El arte de olvidar*), an Argentine scientist with a long career in the study of memory. In the rest of this chapter I will discuss different forms of forgetting[2] and describe two contrasting cases: those of Jill Price and Clive Wearing.

In 2000 Jill Price contacted one of the world's foremost experts on the study of memory, Professor James McGaugh of the University of California, Irvine:

I am thirty-four years old and since I was eleven I have had this unbelievable ability to recall my past. . . . Whenever I see a date flash on television (or anywhere else for that matter) I automatically go back to that day and remember where I was, what I was doing, what day it fell on and on and on and on and on. It is non-stop, uncontrollable and totally exhausting.

2. A more rigorous description of the different forms of forgetting appears in Daniel L. Schacter, *The Seven Sins of Memory: How the Mind Forgets and Remembers* (Boston: Houghton Mifflin, 2001).

McGaugh examined Price with as much skepticism as Luria felt when he first interviewed Shereshevskii, but he quickly found something that made her case substantially different from those described previously: unlike Shereshevskii or savants, Price did not have a special talent for remembering letters or numbers; her memory was very personal, linked to her life and everything related to it. From the age of 10 she had kept a (50,000-page) journal; like Shereshevskii, she justified her obsession with recording everything as a way to try to take the memories out of her head. Thanks to this journal, McGaugh could verify all the facts that Price had told him about her life, and her case became widely known after McGaugh and his colleagues published a paper about it.[3] Price was a guest on several television shows, where she was repeatedly put to the test with different dates and events. During one of those interviews, with ABC's Diane Sawyer, she made what seemed to be her first mistake: she said Grace Kelly died on September 14, 1982, while the book used by the network to check her responses said it had happened on the 10th. Confused at first, Price thought she had been asked something else, but later she insisted that the date she gave was correct. And she was right: the producers checked again and found that the book was wrong.

In spite of her amazing memory, Jill Price did not do well at school. The problem is that her memory is limited to her biography and the events around it; apart from that, she cannot remember a poem, an equation, a foreign language, or any historical event not

3. Elizabeth S. Parker, Larry Cahill, and James L. McGaugh, "A Case of Unusual Autobiographical Remembering," *Neurocase* 12 (2006): 35–49.

directly related to her personal life (including, for example, everything that happened before she was born). Having seen the cases of Funes, Shereshevskii, and the savants, we should not be surprised that, as McGaugh concluded, Price has problems with any test that involves abstract reasoning. In particular, she tends to respond with details instead of more abstract concepts. McGaugh claims that her memory, disproportionate and hyperfocused on her person, may stem from an obsessive compulsive disorder that crowds out any other thoughts:

Most have called [my memory] a gift but I call it a burden. I run my entire life through my head every day and it drives me crazy!!! . . . I think about the past all the time. . . . It's like a running movie that never stops. It's like a split screen. I'll be talking to someone and seeing something else. . . . Like we're sitting here talking and I'm talking to you and in my head I'm thinking about something that happened to me in December 1982, December 17, 1982, it was a Friday, I started to work at Gs (a store) . . .[4]

Just like Claire Tourneur (Solveig Dommartin), who cries inconsolably when seeing her childhood memories unfold on a portable screen in *Until the End of the World*, Jill Price confesses what a pain and a burden it is not to be able to forget even a single detail from every unpleasant situation she has lived through. In the interview with Diane Sawyer she says:

J.P. I still feel bad about stuff that happened thirty years ago.
D.S. But we all do . . . I mean, we can all remember one or two things . . .

4. Ibid.

J.P.   It's not one or two things . . . everything! And I really, like, live it and feel it. And I think, if I'd just done this then I wouldn't have done this and I wouldn't be here, and I'm always, constantly doing that.

Her mother recalls having made a couple of remarks when she became overweight as a teenager. Jill, however, remembers the date and time of each of those remarks (more than 500, according to her). She confesses that having so many memories is intolerable and feels that she should be in a mental hospital. She remembers and relives the death of her husband or the breakups she had with other partners as though they had just happened. This highlights the importance of forgetting in our lives. In my case, I don't really want to relive in detail that day at the beach; I'd rather have it as it is, a faint memory, a little sad, with the nostalgia of a tango. I don't want, like Claire Tourneur, to see my childhood memories played back on a screen.

In a previous chapter we discussed the reconsolidation process: memories are brought back again and again (especially during sleep) and through this repetition they set in the cerebral cortex. But unconsciously our brains use various tricks to ensure that we do not end up like Jill Price. This is because reconsolidation is an active process that goes far beyond a mere recapitulation of established memories. When we bring back a memory, we do not just watch it like a film; every time we relive a past memory we generate a reconstruction that differs from the original. Unconsciously we tend to forget some facts and make others up so that the memory is more pleasant and coherent (and therefore easier to remember). We even enrich and modify these reconstructions based on other information we process as we recall. If I discuss a story by Kafka with someone else, my memory of the story is enriched by the conversation. During the reconsolidation process we can also alter the original memories. For example, if I recall the plot of *The Metamorphosis* and remember that Gregor Samsa wakes transformed into a gigantic cockroach, and then someone more knowledgeable tells me that in truth Kafka never makes it explicit that Samsa becomes a cockroach but just uses the vague German word *Ungeziefer* (vermin, insect), I will be able to correct my memory and recall the correct term next time.

The cliché has it that all times past were better, but this may just reflect the healthy filtering—that is, the appeasement of the least pleasant parts of our memories—that takes place during the reconsolidation process. We recall with nostalgia our childhood in elementary school, but we tend to forget how excruciating it was to wake up early every day, study for a test, or sit through one class after another. In a famous experiment,[5] Elizabeth Loftus showed how memory reconsolidation can lead to fantasy or to the formation of false memories. Loftus showed subjects films of automobile accidents and then asked some of them how fast the cars were going when they *hit* each other. She asked the same question to another group, except that she used the word *collided*. A third group had to estimate the cars' speeds when they *smashed* into each other, with a fourth group she used the word *contacted*, and with the fifth she used *bumped*. The subjects who saw *smashed* gave the largest speed estimates (40.5 mph on average), followed by those who saw

5. Elizabeth F. Loftus and John C. Palmer, "Reconstruction of Automobile Destruction: An Example of the Interaction between Language and Memory," *Journal of Verbal Learning and Verbal Behavior* 13 (1974): 585–589.

collided (39.3 mph), *bumped* (38.1 mph), *hit* (34.0 mph), and finally *contacted* (31.8 mph). Even more surprising was the fact that, a week later, Loftus asked the same subjects if they had seen broken glass in the accident (there was none). Thirty-two percent of the subjects that saw the word *smashed* answered they had, while only 14 percent of those who saw *hit* did. Loftus's results show how fragile our memories are and how they are prone to manipulation during the reconsolidation process—all it took was changing a single word in one question. Beyond their scientific interest, these discoveries were of enormous practical importance because they highlighted the subjectivity of eyewitnesses at trials and how their testimony can be manipulated by the way in which different questions are asked.[6]

Héctor Maldonado, one of the pioneers of memory research in Argentina, starts his book *La memoria animal* with the legend of Mnemosyne, the personification of memory in Greek mythology (and mother of the nine muses, which implies, according to Maldonado, that artistic inspiration stems from memory), and Lethe, the spirit of forgetting. According to legend, Mnemosyne and Lethe were associated with two antagonistic rivers; drinking from Mnemosyne's waters conferred total memory, while drinking from

6. More details can be found in:
- Elizabeth F. Loftus, "Our Changeable Memories: Legal and Practical Implications," *Nature Reviews Neuroscience* 4 (2003): 231–234;
- Elizabeth F. Loftus, "Creating False Memories," *Scientific American* 277 (1997): 70–75;
- Alan D. Baddeley, Michael W. Eysenck, and Michael C. Anderson, *Memory* (Hove, East Sussex, UK, and New York: Psychology Press, 2009).

FIGURE 8.1

Mnemosyne (oil painting by Dante Gabriel Rossetti) and three of the nine Muses (oil painting by Eustache Le Sueur): Clio (history), Euterpe (music), and Thalia (comedy). Photograph of the Lethe river in Alaska, whose name was inspired by the Greek legend.

Lethe (which flowed through the cavern of Hypnos, the personification of sleep) produced complete oblivion.

In one of Plato's dialogs, Critias invokes Mnemosyne before embarking on a lengthy exposition, since the success of his speech largely depends on gaining her favor to be able to remember it. The importance of memory in the ancient world should not be underestimated: at a time when there was no easy way to jot down notes, orators depended exclusively on their memories in delivering their speeches. Quintilian, the Roman orator, writes in his *Institutio oratoria*:

Nesciretur tamen quanta vis esset eius, quanta divinitas illa, nisi [in] hoc lumen orandi extulisset.[7]

[We would have never known how great and divine the power of memory is, were it not for the fact that it is memory that has brought rhetoric to its present glory.]

However, as we have seen, we must resist the temptation to drink from Mnemosyne's waters if we are not to end up like Funes. We should also refrain from drinking from Lethe, whose waters, according to Greek mythology, were drunk at the moment of death so the past life would be forgotten and reincarnation could occur. Unfortunately, Lethe's water is given to some of us before death, either through Alzheimer's disease, Korsakov's syndrome, or in

7. Quoted in Frances A. Yates, *The Art of Memory* (London: Pimlico, 1992). In the same book, Yates mentions Themistocles, the Athenian general who, cunningly, declines to learn the art of memory because he would rather learn the science of forgetting.

rare cases like those of H.M. or Clive Wearing, a patient with an even more profound amnesia.

Clive Wearing was a brilliant musician, the producer of a music show for the BBC, and a leading expert on the works of Renaissance composer Orlandus Lassus. At the peak of his career, on March 29, 1985, he suffered from a viral encephalitis that destroyed a large portion of his brain, including the hippocampus, the temporal lobes, and part of the frontal lobes. Wearing survived, in spite of his doctors' expectations, but paid a very high price: not only was he unable to form new memories, like H.M., but also his memory of the past was almost completely erased. His memories can be counted on the fingers of one hand: he remembers his wife; he knows he has children from a previous marriage but does not remember their names or even how many they are; he knows he is a musician but does not remember going to a single concert; he knows he attended Cambridge. He cannot remember even the basic things we all take for granted. His wife says he once ate a lemon with the rind; he cannot recognize the toothpaste in his bathroom; he mixes up soap and talc; he once thought that an umbrella was a scarf; he cannot distinguish between his wife's sweaters and his shirts.[8] Clive Wearing cannot remember the sentence he just spoke, hold a conversation, read, or watch television. He spends his days as though every half-minute he were waking up from a prolonged

8. Barbara Wilson and Deborah Wearing, "Prisoner of Consciousness: A State of Just Awakening Following Herpes Simplex Encephalitis," in *Broken Memories: Case Studies in Memory Impairment*, ed. Ruth Campbell and Martin Conway (Oxford: Blackwell, 1995). Wearing's case is also discussed by Oliver Sacks in *Musicophilia* (New York: Knopf, 2007).

coma. Every time he sees his wife he hugs and kisses her, moved to tears, thinking he has not seen her in years. The only thing he can still do is play the piano.

Clive Wearing desperately tried to understand what was happening to him, but could not. His wife tells that shortly after the onset of amnesia, when he realized that he was completely incapable of thinking, he cried nonstop for a couple of months. At that point he tried to start a journal and record everything he did; but the notes he took made no sense to anyone, and least of all to him, who was feeling all the time as if he had just awoken. This is an excerpt from Wearing's journal for Thursday, August 30, 1990:

2:50 A.M.: I almost wake up for the first time. Patience. [The word "almost" is added above the line.]

7:36 A.M.: I do wake for the first real time for ?6yrs. Patience. One of my records is distantly audible.

7:55 A.M.: Now I am totally awake (1st real time). I look for b'fast. [The word "totally" and the phrase "1st real time" are crossed out.]

8:16 A.M.: Now I am perfectly awake (1st real time). I wait for b'fast. [The word "perfectly" and the phrase "1st real time" are crossed out.]

8:44 A.M.: Now I am superlatively awake (1st real time). Patience helps. [The word "superlatively" and the phrase "1st real time" are crossed out.]

9:58 A.M.: Now I am really completely awake (1st actual time). [The phrases "really completely" and "1st actual time" are crossed out.]

While recording the 7:36 entry he probably added the word "almost" to the 2:50 note (he was almost awake before, but now he is indeed awake). At 7:55 he says he is "totally" awake so that the 7:36 entry—when he just awoke—will make sense. At 8:16 he probably crossed out the word "totally" from the 7:55 entry because he thought he could not have awoken 20 minutes earlier; at 8:44 he may have crossed out "perfectly" from the 8:16 entry, and so on. The sad fact is that Clive Wearing could not make sense of what he had written just minutes before, and this made him correct himself continually.

On a BBC documentary, Wearing's wife says that he lives constantly in "a moment to moment consciousness," an isolated present "with no past to anchor it and no future to look ahead to." At a restaurant he cannot recognize the cooked vegetables on his plate, calling them a salad (a cooked salad, as he corrects himself after he tries it). When he talks to his wife, she repeats herself again and again, every few seconds, but he cannot remember anything. Sipping tea, and later sitting on a couch in his room, Wearing says:

I've never seen anyone at all . . . I never heard a word until now. I know I haven't dreamed, even—day and night are the same, blank. Precisely like death. No thoughts at all. Brain has been inactive, and day and night exactly the same. No dreams, even. . . . It's been like death. I've never seen a human being before, never had a dream or a thought. Brain has been totally inactive, day and night the same, no thoughts at all. As far as I'm concerned, the doctors have been totally incompetent. I've never seen a doctor the whole time.[9]

He repeats himself again and again, always failing to remember what he just said. His wife comes back and he jumps from

9. "The Man with a 30 Second Memory," available at http://www.youtube.com/watch?v=WmzU47i2xgw.

the couch and, movingly, embraces her with passion. Then they dance and kiss on camera, as if they had not seen each other for years.[10]

10. In his books *The Man Who Mistook His Wife for a Hat* and *An Anthropologist on Mars*, Oliver Sacks describes many cases of patients with profound amnesia. One of these, "The Lost Mariner," tells the story of a man with Korsakov's syndrome (due to chronic alcoholism) who could not remember anything that happened after 1945. Sacks describes the horror felt by this man when seeing his face in the mirror and his confusion when seeing a photograph of the earth taken from the moon, something inconceivable for a mind stuck in 1945.

PERCEPTION AND MEMORY

One of the most fundamental questions in philosophy involves the existence of the being and the external world. It is as simple and yet as profound as asking, Do I exist? Or, Does what I see around me exist?

In book VII of his *Republic*, Plato proposed the famous allegory of the cave to posit the existence of "universals": Imagine a group of people living in chains in a cave so that they can see only shadows of what happens behind them; to these people, the shadows become their reality because they are unable to perceive what generates them. Likewise, Plato argued that we see only the manifestation of universal concepts because the concepts themselves are beyond our perception. For example, if we see an apple, we perceive an instance of the concept "apple," which exists independently of both the particular apple and those who perceive it (i.e., us). The

FIGURE 9.1
A detail of *Plato's Allegory of the Cave*, a 1604 engraving by Jan Saenredam. The people in the cavern (at the bottom) can perceive only the shadows projected on the wall (their reality) and not what generates them (the universal reality).

philosopher's mission is then to abstract and conceive these universals starting from the shadows that he or she sees.

This philosophical doctrine, dubbed "Platonic realism," was later criticized by Aristotle, Plato's most famous disciple, who saw no need to distinguish between the world of ideas—of universals—and the world of things. According to Aristotle, ideas are nothing more than abstractions we make of ordinary things, a position that was later pursued further by Thomas Aquinas. The final break with realism, however, was to come two millennia later with René

Descartes, the father of modern philosophy. If we had to summarize Descartes's thought in one sentence, it would certainly be "Cogito ergo sum" ("I think, therefore I am"). To arrive at fundamental, unquestionable truths, Descartes systematically put everything in doubt and thus tested the fundamental tenets of realism. He concluded that our senses often fool us and that we have no way of ruling out the possibility that what we see around us is part of a dream conceived by an evil and powerful genius. But this genius, no matter how sophisticated its ruse, cannot make us doubt our own existence, because we are at least capable of having a doubt. Then our ability to think is, according to Descartes, what assures us that we exist. This assessment of the value of thought led him to postulate that the mind (the essence of being) is separate from the body (and a dubious reality), which constitutes the basis of Cartesian dualism.

Now, what does this philosophical discussion have to do with perception? And further, what does perception have to do with memory? As we shall see, a lot. . . . Descartes laid the foundations of "idealism":[1] there may not be an external reality, as Platonic

---

1. I am speaking of idealism in a broad sense. Descartes is in fact regarded, along with Baruch Spinoza and Gottfried von Leibniz, as one of the paragons of rationalism, which postulates that knowledge is obtained only through reason (whether by intuition or deduction) and independently of sensory experience. The classic example of rationalist philosophy is trigonometry, a construction founded on intuitive axioms and proofs that elaborate on them. At the other extreme, empiricists thought that knowledge comes about only through experience and the senses. These two schools of thought later merged in Kant's transcendental idealism, according to which we get to know only representations of things that we create

realism propounds; rather any knowledge of the external world resides in the mind. Descartes's ideas were elaborated upon by the British empiricists, John Locke, David Hume, and George, Bishop Berkeley—for whom the mind is a sort of *tabula rasa,* an empty slate on which we constantly inscribe the knowledge that we accrue through life experience and sensory perception—and by Immanuel Kant, who argued that we only get to know the representations we make of things rather than the things themselves (*die Dinge an sich*).

These ideas have not been alien to neuroscience or to Borges, who discusses the British empiricists in several stories. Ultimately, the reality we experience is given by a series of perceptions that take place in our brains. Aristotle proposed that the psyche uses the information that we gather with our senses to generate the ideas, concepts, and abstractions that comprise our thought. He writes in *On the Soul* (427b, 428b):

Thinking is different from perceiving and is held to be in part imagination, in part judgement. . . . To imagine is therefore (on this view) identical with the thinking of exactly the same as what one in the strictest sense perceives.[2]

in our minds starting from our innate intuition of space and time (in other words, we conceive things at a time and a place). According to Kant, knowledge encompasses both the senses—so we can be aware of things—and reason—so we can conceptualize things and think about them.

2. Aristotle, *De anima* (*On the Soul*), trans. J. A. Smith, in *The Basic Works of Aristotle*, ed. Richard McKeon (New York: Random House, 1941), pp. 587, 588.

In other words, Aristotle understood the importance of the psyche in perceiving external reality. Later, the idealists, aware of how easily we can be "deceived" by our senses, went further and doubted reality itself. These "deceptions," however, are nothing but the processes carried out by neurons—call them brain, mind, or psyche—to comprehend the information provided by the senses. Thus there is an enormous difference between *sensation*, the visual stimulus that strikes neurons in the retina, and *perception*, the meaning we assign to that stimulus, be it a house, a face, or a Bengal tiger. Absurd as it may seem, we do not see with our eyes or hear with our ears; we actually see and hear with our brains.

Along the same lines, in what seems like an argument for idealism or, if you will, a late-twentieth-century reinterpretation of Berkeley's thought, Francis Crick writes:

What you see is not what is *really* there; it is what your brain *believes* is there. . . . Seeing is an active, constructive process. Your brain makes the best interpretation it can according to its previous experience and the limited and ambiguous information provided by our eyes.[3]

The difference between sensation and perception had already been noted, albeit in other words, by Ptolemy—the distinguished Egyptian astronomer who reasoned that our conception of the size of an object depends not only on the angle from which we see it but also on our estimate of how far it is from us—and by Aristotle himself, who states the following in *On the Soul* (428b):

3. F. H. C. Crick, *The Astonishing Hypothesis* (New York: Simon and Schuster, 1993), p. 31. The italics are Crick's.

But what we imagine is sometimes false though our contemporaneous judgement about it is true; e.g. we imagine the sun to be a foot in diameter, though we are convinced that it is larger than the inhabited part of the earth.[4]

This observation was pursued in greater detail by Alhazen (or Ibn al-Haytham, the medieval Islamic scientist considered to be the father of optics), who deduced that the perception of size depends on assessments we make unconsciously. Alhazen also noticed that something similar happens with color perception, since we tend to

FIGURE 9.2
Left: Image of Alhazen (Ibn al-Haytham, 965–1040) used in Iraqi currency.
Right: Hermann von Helmholtz (1821–1894).

4. Aristotle, *De anima* (*On the Soul*), p. 588.

see that an object has the same color regardless of the way it is illuminated. Similar observations were made by Descartes in his treatise on optics, but it was the German physicist and physician Hermann von Helmholtz who developed this idea more fully at the end of the nineteenth century.

Few scientists in history have been as versatile as Helmholtz, who among many other achievements formulated the principle of conservation of energy and the concept of free energy in thermodynamics; invented the ophthalmoscope to examine the retina; measured the speed at which signals travel in nerves; created the modern theory of colors by using three variables—hue, saturation, and brightness—to characterize them and proposed that any color can be generated by mixing three basic ones; introduced a mathematical formulation of acoustic vibrations and built resonators from empty bottles; and so on.

Helmholtz observed that the information gathered by our eyes (as well as by our senses of hearing, touch, smell, and taste) is very meager and therefore the brain must make *unconscious inferences* in order to interpret external reality. In other words, the brain copes with incomplete data by making predictions based on past experience, an idea very much akin to the line of thought of the British empiricists. For example, let us imagine that, with eyes closed, we feel the texture of a coin with our thumb and index finger. If our fingers are separated, forming the shape of an "L", we will have no doubt that we are touching two different coins, whereas if the fingers are nearly touching and facing each other, we will conclude that it is one and the same coin. Interestingly, the sensory information that we have (gained only through touch) is exactly the same in both cases, but unconsciously we ascribe our sensation to the presence of one or two coins depending on how our fingers are arranged. This is one of the many unconscious inferences the brain makes to attribute meaning to the information that comes through the senses.

According to Helmholtz, what we see are not copies but *signs* of external objects. In fact, these signs do not have to resemble the objects they represent, so long as they are reproducible—i.e., so long as a given object evokes the same sign in different circumstances. In *The Facts of Perception* (1878) Helmholtz writes (with quite an idealistic outlook):

The objects in the space around us appear to possess the qualities of our sensations. They appear to be red or green, cold or warm, to have an odor or a taste, and so on. Yet these qualities of sensations belong only to our nervous system and do not extend at all into the space around us.[5]

Perhaps the strongest argument supporting Helmholtz's view is given by the existence of visual illusions, which are typically due to inferences made by the brain. Figure 9.3 shows three classic illusions. The one on the left is Kanizsa's triangle, in which we infer the shape of a white triangle from looking at its angles—the triangle disappears completely if we even slightly misalign the wedges—and the interrupted outline of the black occluded triangle. The one in the center is Ponzo's illusion: the two horizontal lines have the same length, but the one on top appears to be longer because the perspective suggested by the oblique lines makes it look farther away.

5. Hermann von Helmholtz, "The Facts of Perception," in *Selected Writings of Hermann von Helmholtz*, ed. Russell Kahl (Middletown, CT: Wesleyan University Press, 1971).

FIGURE 9.3
Three visual illusions: Kanizsa's triangle, Ponzo's illusion, and an illusion of relief.

The illustration on the right shows two circles that appear to have depth, the one at left seemingly pushing into the sheet and the one at right apparently sticking out. In this last case the illusion arises because we unconsciously assume the circles to be illuminated from above; the light reflecting off the bottom rim of the circle gives the illusion of depth and the one bouncing off the top suggests protrusion. It turns out the circles are identical but rotated 180 degrees (just turn the book upside down to upend the illusion).

The inferences described by Helmholtz are what allow us to perceive reality, to make sense of what we see. They also help us to extract and bind features so that we see a given object as a whole. The rules by which we perform this binding process were studied by the Gestalt psychologists, a school of psychology developed in Germany in the 1930s which postulates that objects are grouped based on continuity, similarity, proximity, and other properties. Kanizsa's triangle, for example, is perceived by aggregating three angles into a single object, while Ponzo's illusion is based on the sensation of perspective that stems from considering the slanting lines as one object (like a road or railway tracks).

Let us now return to the question we have left pending. What is the relation between perception and memory? The answer is very simple: we cannot recognize anything (be it a face, a dog, a spoon) if we do not have a previous memory of it. When I recognize someone on the street, I contrast the visual image that enters my eyes with the memory I have of that person. When I see a tree, I abstract its salient features and compare them to those stored in my memory of what trees look like. It is for this reason that we said in previous chapters that memories might be located in the same places where perception is carried out with the respective senses: the visual memory of a face would be located in the inferior temporal cortex, where face recognition is processed, while the auditory memory of a passing train would be located in the region of the temporal cortex that deals with sounds.

One of the most famous cases described by writer and neurologist Oliver Sacks is that of Dr. P., a talented musician whose life story provides the title for Sacks's *The Man Who Mistook His Wife for a Hat*. Dr. P. was unable to recognize his students' faces, but could identify them by their voices as soon as they spoke. He was incapable of recognizing his colleagues, relatives, and even pictures of himself (he only recognized Einstein because of his well-known mustache and hairstyle, and also his brother, a man with a square jaw and big teeth). Sacks relates that, on their first encounter, Dr. P. was unable to recognize the shoe he had just taken off for a routine exam, and that, incredible as it sounds, he mistook his wife's head for a hat. Dr. P. suffered from visual agnosia—the failure to recognize objects despite having no primary visual deficits (he could perceive contours, colors, depth, etc.). This deficit in perception did not mean that Dr. P. could not recognize or name objects at all (as in cases of aphasia). In fact, patients with visual agnosia usually have no problem recognizing an object once they touch it, or identifying people when they hear their voices. The deficit is limited to visual perception.

There are two main types of agnosia caused by two different types of brain lesions: perceptive agnosia and associative agnosia. Patients with perceptive agnosia cannot see an object as a whole: they perceive only details (like Einstein's mane) but cannot combine them to generate the "Gestalt," the gist, the essence of the figure. On the other hand, patients with associative agnosia perceive objects but cannot associate a meaning to them. In other words, they cannot contrast an object with the memory they have of it, and for that reason they cannot recognize it. They have no problem copying a picture—their visual perception is intact—but they cannot say what it is. Cases of associative agnosia are thus clear evidence of the substantial relationship between perception and memory.

In previous chapters we saw how we discard a huge amount of information and end up storing only a minute number of facts. We also discussed the importance of forgetting. Most of the little we store in our brains will get diluted in time. But the biggest information loss takes place even before the first memories are formed, and it stems from the nature of perception. As Helmholtz argued, we process "signs" of what we see, not copies. We generate abstractions, concepts, Platonic universals; we give a meaning to what strikes our retinas and discard countless details. In a given instant we see only a tiny fraction of what is in front of us; our eyes are constantly moving and our perception of the world around us as a whole is an illusion created by the brain. We need only general features to be able to infer the rest.

We already said that incorrect inferences generate visual illusions. These inferences are so entrenched in our brains that we cannot avoid them even if we understand how they work; we cannot help perceiving the top line in Ponzo's illusion as being longer even if we are well aware that both lines are equal. We can show just how little information about the external world we process by considering the phenomenon of "change blindness." Two seemingly identical images are switched back and forth on a computer screen and the task is to find the difference between the two. Most subjects usually take several seconds to find the difference and are quite surprised at not having found it earlier, given that it is so obvious in hindsight. This can be explained once we realize that, even though we think we are seeing everything that lies in front of us, in truth we observe only general features. A similar principle is at work in the "Spot the differences" games that appear in newspapers. Figure 9.4 shows two versions of part of Raphael's famous fresco *The School of Athens*, one of them with a single altered detail. The reader is encouraged to take a few seconds to spot the difference before reading on.

The fresco shows Plato and Aristotle at the center, surrounded by other philosophers and students of the Athenian school.[6] The one at far left is Socrates, Plato's mentor and the protagonist of his dialogs. In "Keats's Nightingale" (from 1952's *Other Inquisitions*),

---

6. In fact there was no single "school of Athens"; Plato taught at a private school called the Academy, while Aristotle started his own school, free and open to the public, called the Lyceum.

FIGURE 9.4

Part of *The School of Athens* by Raphael, and the same view with an altered detail.

Borges quotes Coleridge, who once observed that every man is born an Aristotelian or a Platonist. The first has his feet on the ground, believes in the reality of things, while the second seeks truth in universals, in what goes beyond the particular instances exhibited by the senses. Raphael shows this difference clearly in his original fresco (figure 9.4, top): while Plato points upward, to the world of ideas, Aristotle, in an eloquent gesture, keeps his focus down to earth. This is actually the altered detail in the copy, a bit of philosophical whimsy in which Plato, with a gesture of his right hand, accepts his disciple's point of view. But this "metaphysical heresy" is nothing more than a game that illustrates the principle I have just described. It generally takes us several seconds to spot the difference: we think we see the painting as a whole, and it surprises us to find the detail whose alteration eludes us. This is normal and inevitable. We process only a minimal fraction of the information that passes through our eyes, the fraction we care about. We use this information to extract signs, and we create concepts and form internal representations that are the basis of our thought; we then form memories and keep only those that are most relevant, draining the rest into Lethe, directly into the realm of oblivion.

NEUROPHYSIOLOGY OF VISION

In this chapter I will temporarily leave Borges aside in order to describe in more detail the process of perception and the way in which visual stimuli are encoded by neurons in different areas of the brain. This will then lead me to a description of the neurons involved in the generation of memories, whose interpretation is closely related to the ideas developed by Borges in "Funes the Memorious."

At the end of the nineteenth century it was believed that the brain was composed of a continuous mesh of intertwined nerve cells. This so-called "reticular theory" was far from outlandish, considering that under the microscope the brain looked like an enormously complex reticulum (network) of connections forming a mesh of filaments with neurons scattered throughout. One of the most renowned advocates of the reticular theory was Camillo Golgi (1843–1926), who invented a method to visualize neurons by staining them. He was later proved wrong by Santiago Ramón y Cajal, ironically using the very technique that Golgi had developed. Like other scientists of his era, Golgi had the problem that the reticula observed through the microscope were of such complexity that it was impossible to single out the connections pertaining to each neuron. Cajal's genius consisted in using tissue from newborn animals, whose reticula were not fully developed and thus were easier to visualize. The use of immature animals, Golgi's stains, and an admirable will that kept him for hours at the microscope led Cajal to demonstrate the individuality of neurons and develop a "neuron doctrine," arguing that neurons are the fundamental atoms of cerebral activity.[1] This discovery, the foundation of modern neuroscience, won Cajal the 1906 Nobel Prize in Physiology or Medicine (shared, curiously, with Golgi, who advocated for the opposite theory).

Cajal, whom many consider to be the father of neuroscience and the greatest Spanish scientist who ever lived, wanted to be a painter when he was a child. His talent and vocation for the arts can be seen in the hundreds of drawings he made of neuron meshes in different areas of the brain. Most of these drawings are kept in the Museo Cajal, in Madrid, but they could just as well be exhibited in a museum of modern art (and indeed their reproductions adorn many neuroscience centers). Amusingly, if we were to remove the microscopes from the photo of Cajal, he would somehow look like a painter in his atelier. I cannot help noticing the simplicity of his laboratory: a table, a chair, a shelf with some chemicals, and a pair

1. Neurons have incoming and outgoing connections (dendrites and axons) which generate a one-way flow of information; they receive information through the dendrites and transmit it to other neurons through the axons.

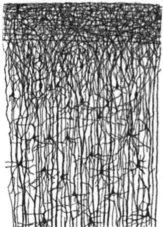
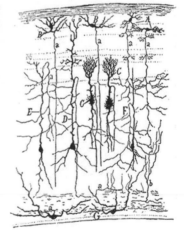

FIGURE 10.1
Santiago Ramón y Cajal (1852–1934) and two of his drawings: the one on the left shows the structure of a neural network in the cerebral cortex; the one on the right shows the morphologies of different types of neurons.

of microscopes. It would be very difficult for a nonspecialist to guess that the man in the picture revolutionized neuroscience and won the Nobel Prize.

Besides being an extraordinary scientist, Cajal was a prolific writer and a champion of Spanish science. Worthy of note among his writings is his 1897 induction speech at the Academia de Ciencias Exactas, Físicas y Naturales, later published in *Advice for a Young Investigator*,[2] aimed at young researchers. In this speech, Cajal offers advice on the qualities that scientists should develop, on how to organize and where to publish their results, on the attitude they should show toward celebrated researchers, and even on the type of woman that a male scientist should marry.[3]

2. Santiago Ramón y Cajal, *Advice for a Young Investigator* (*Reglas y consejos sobre investigación científica*), trans. Neely Swanson and Larry W. Swanson (Cambridge, MA: MIT Press, 1999).

3. Cajal states that, though science and love must be kept separate, the scientist should marry, since unmarried men suffer more from sexual preoccupation. He goes on to evaluate different kinds of women and assess how convenient they would be to a man of science. According to Cajal, the wife of a scientist can be:

• Intellectual: Quite commendable, since she can help with scientific endeavors, but rare, especially in Spain.
• Opulent: Rich, but prone to vice.
• Artistic: Needy, dramatic, complicated.
• Thrifty and industrious: Simple, of good character, physically and mentally healthy. Will be useful to the scientist because she can provide support, and she is also quite common in Spanish society.

This enumeration, though perhaps not appropriate nowadays, gives an idea of Cajal's eye for detail and his concern for guiding young minds (and, perhaps, his sense of humor).

Since Cajal we know that the brain is composed of trillions of neurons, which are the elementary units of thought (just as atoms are of matter). Simplifying things a bit, we can state that neurons are either silent or firing what are known as "action potentials."[4] Given these possible states, a single neuron can encode only two options (it may be either silent or active),[5] two neurons can encode up to four (silent-silent; silent-active; active-silent; active-active), and so on. An increasing number of neurons can encode an increasingly complex amount of information, and, in fact, one of the most interesting open problems in current neuroscience is to understand how millions of neurons fire in different ways, either to perceive and understand the world around us—to see, hear, feel—or to create new memories, new thoughts, or the consciousness we have of ourselves. As Francis Crick eloquently put it at the beginning of his book *The Astonishing Hypothesis*, everything that we are and do is nothing more than the behavior of a large number of neurons. No one doubts this; the big question is how. Neuroscientists keep looking for a Rosetta stone that will lead us to understand how from a pattern of neuronal activity we generate our sense of self and all its accompanying functions.

4. In making this simplification I am mainly leaving aside the neuronal activity below the triggering threshold.

5. An active neuron is one that increases its firing frequency, for example from 1 spike per second to 10 spikes per second. This statement hides another simplification, since we are assuming that each neuron encodes information only through its firing frequency (though it could also transmit information based on the precise timing of each action potential).

Let us go back to vision, the subject of the previous chapter, but with a different focus. The process of visual perception begins in the retina, with neurons that respond to the intensity of light at a given point.[6] We can understand this idea by analogy with a television screen: it is as though each retinal neuron encodes the light intensity at a single pixel; each neuron encodes a very limited and local amount of information, but if we aggregate the activities of all of them (all neurons, or all pixels) we can reconstruct the image on the screen. This information is transmitted—through the optical nerve and the lateral geniculate nucleus of the thalamus—to the primary visual cortex, which has some 200 neurons for each neuron in the

---

6. Some of these neurons respond to points of light (exhibiting "on responses") and others to points of dark ("off responses"). One type of neurons, called "cones," respond to different frequencies of the visible spectrum (corresponding to red, green, and blue light), allowing us to perceive color. To simplify, I am not considering neurons with more complex responses in the retina, and center-surround organization: a neuron with on responses at a given location has off responses in the surroundings (and vice versa). So, rather than local intensity, it is more accurate to say that retinal neurons fire to local contrast. Neuronal responses can be measured by inserting microelectrodes (filaments much thinner than a human hair) into the brain. These microelectrodes behave somewhat like microphones that allow us to listen to the neural activity taking place in their surroundings. The signals they pick up can in fact be processed and played on speakers: one hears a background noise, similar to that made by a television set with its antenna and cable box disconnected, and on top of that the firing of the neurons, something like: shhhhh . . . tac, shhhhh . . . tac tac. . . . When a neuron is silent, it fires every once in a while, but when it becomes active it starts firing like a machine gun: shhhhh tac tac tac tac tac. . . .

retina. We have already remarked that we fail to perceive much of what is in front of us. Then why do we have so many neurons for such a poor visual representation? Because the brain, rather than creating a representation, a copy of what we see, makes inferences so we can extract its meaning. A high-definition television set gives us a very precise representation of an image, but, no matter how sophisticated its technology, it fails to inform us if what we are watching is a soap opera, an action film, a picture of a landscape, or an animal documentary. This is, on the other hand, something we can do effortlessly with the powerful machinery lodged in our visual cortex, even if we do not have the resolution of a high-definition television or, for that matter, of a cell phone digital camera.

The fact that there are so many neurons in the primary visual cortex simply reaffirms what Helmholtz stated in the nineteenth century: the brain does not reproduce visual stimuli like a digital camera or a television, but rather processes their meaning starting from relatively little information and a set of unconscious inferences. Now, if the neurons in the primary visual cortex do not simply copy the information detected by the retina, what do they do? This was what David Hubel and Torsten Wiesel at the Johns Hopkins University set out to study in the late 1950s. Following a serendipitous event[7] and a spectacular series of experiments that

---

7. In their initial experiments, Hubel and Wiesel showed visual stimuli to anesthetized cats using an ophthalmoscope that blocked other light from entering the cats' eyes. In the ophthalmoscope they placed opaque slides with holes of different sizes so as to create images of shiny dots in different spots. Other slides were made of transparent glass and had a metal dot adhered so as to generate images of dark dots. A few weeks after

ensued (which earned them the Nobel Prize in Physiology or Medicine in 1981), Hubel and Wiesel discovered neurons in the primary visual cortex that respond to oriented lines.[8] This information eventually reaches the inferior temporal lobe (figure 10.2), where "face cells"—i.e., neurons that respond to human or monkey faces but not to other images, for example of hands, fruits, or houses—were found in experiments with monkeys.[9]

Now, how do we go from recognizing points in the retina to recognizing more complex objects in higher brain areas? Figure 10.3 illustrates this transformation. First, each neuron in the retina

they started these experiments they noticed that a neuron fired when they put one of the glass slides in the ophthalmoscope; but the neuron's response had nothing to do with the dark dot and was triggered rather by the edge of the slide (actually a straight line with a given orientation) as it was put in or taken out. It would be unfair to consider Hubel and Wiesel's discovery as merely a lucky break. After all, they knew how neuronal responses happen in the retina (mainly through experiments carried out by Stephen Kuffler, their mentor) and were perceptive enough to study the primary visual cortex, an area completely unexplored till then. They also had the perseverance to record and characterize the response from this first neuron for nine straight hours; many others would have given up after much less time and would have dismissed that response as an artifact.

8. David H. Hubel and Torsten N. Wiesel, "Receptive Fields of Single Neurones in the Cat's Striate Cortex," *Journal of Physiology* 148 (1959): 574–591.

9. Charles G. Gross, David B. Bender, and Carlos Eduardo Rocha-Miranda, "Visual Receptive Fields of Neurons in Inferotemporal Cortex of the Monkey," *Science* 166 (1969): 1303–1306.

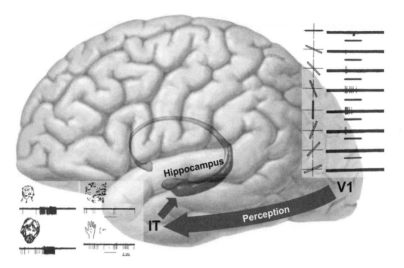

FIGURE 10.2

Visual processing: Neurons in the primary visual cortex (located in the rear of the brain and denoted by V1) respond to lines of a given orientation (a vertical line, in this case). This information is transmitted to the higher visual areas through the so-called perceptual, or ventral, pathway and ends in the inferotemporal cortex (IT), where neurons have been found to respond to faces and no other visual stimuli. Information from IT is transmitted directly to the hippocampus.

(left) responds to a particular point, and we can infer the outline of a cube starting from the activity of about thirty of them. Next, the neurons in the primary visual cortex (center) fire in response to oriented lines; fewer neurons are involved and yet the cube is more clearly seen. This information is received in turn by neurons in higher visual areas (right), which are triggered by more complex patterns—for example, the angles defined by the crossing of two or

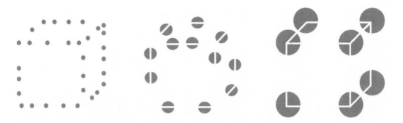

FIGURE 10.3

An example of the types of representations given by the neurons in the retina (left), the primary visual cortex (center), and the higher visual areas (right). Each neuron responds at a given place, in what is known as its receptive field; the neurons in the higher visual areas respond to more elaborate visual patterns (and have larger receptive fields), thus forming a clearer image with fewer neurons.

three lines. Even fewer neurons are needed and the image is more evident.[10]

As the processing of visual information progresses through different brain areas, the information represented by each neuron becomes more complex, and at the same time fewer neurons are needed to encode a given stimulus. In the late 1960s, Polish psychologist Jerzy Konorski wondered if at the end of this process there might be individual neurons that represent an object or a person

10. Further details on the representations formed by neurons in the different areas dedicated to visual processing can be found in:
•   Nikos Logothetis and David Sheinberg, "Visual Object Recognition," *Annual Review of Neuroscience* 19 (1996): 577–621;
•   Kenji Tanaka, "Inferotemporal Cortex and Object Vision," *Annual Review of Neuroscience* 19 (1996): 109–139.

as a whole.[11] Is there, for example, a neuron that represents the concept of "my home," another one that responds to "my dog," and another for "my grandmother"? This idea is in fact much older. At the end of the nineteenth century, William James proposed the following:

> Every brain-cell has its own individual consciousness, which no other cell knows anything about, all individual consciousnesses being 'ejective' to each other. There is, however, among the cells one central or pontifical one to which *our* consciousness is attached. But the events of all the other cells physically influence this arch-cell; and through producing their joint effects on it, these other cells may be said to 'combine.'[12]

The "pontifical" cell to which James refers would be an individual neuron that receives information from others and decides, for example, whether what I am seeing is the face of a particular person I know or not. (This idea was later taken up by Sir Charles Sherrington, who observed that, rather than having such pontifical neurons, the brain makes decisions based on a "millionfold democracy" in which each neuron codifies a different detail of what we see.)[13] This type of representation, given by what Konorski called

11. Jerzy Konorski, *Integrative Activity of the Brain: An Interdisciplinary Approach* (Chicago: University of Chicago Press, 1967).

12. William James, *The Principles of Psychology*, authorized ed., vol. 1 (New York: Henry Holt, 1890; repr., New York: Dover, 1950), p. 179. It is incredible that James made this argument before Cajal proposed his single-neuron doctrine. (The italics in the quote are James's.)

13. Sir Charles Sherrington, *Man on His Nature* (New York: Macmillan, 1940; 2d ed., Cambridge: Cambridge University Press, 1953), p. 277.

gnostic neurons, is popularly known as *grandmother cell coding* due to a famous parable told by Jerry Lettvin in a class taught at MIT.[14] Whether grandmother cells exist or not (either as single neurons or as collections of neurons for each concept) is a question that has given rise to heated arguments in more than one neuroscience conference and will become more relevant as we discuss the results presented in the next chapter. I would also like to mention the contribution of Horace Barlow, who in a seminal 1972 paper[15] refined James's theory by arguing that, instead of a single pontiff, the brain uses a "college of cardinals" to make decisions. Barlow called them "cardinal neurons"; it is what we know today as "sparse coding," a representation given by a relatively small number of neurons (but not by a single one).

14. In a course titled Biological Foundations of Perception and Knowledge taught in 1969, Jerry Lettvin told the (fictitious) story of his cousin, Akakhi Akakhievitch, a great if unknown surgeon from the Urals. Convinced that ideas are stored in specific neurons, Akakhievitch located 18,000 neurons that responded to the concept of "mother," whether seen from the front, from the side, in a caricature, in a photograph, or through an abstraction. At this point he was visited by Portnoy, the character in Philip Roth's novel, who was obsessed with his mother. To cure him, Akakhievitch led him to the operating table and proceeded to ablate each and every one of those cells. After the surgery, Portnoy completely lost the concept of "mother," and Akakhievitch, encouraged by his enormous success, went on to focus on his next endeavor: the grandmother cells. The parable, abridged from a letter Lettvin sent Horace Barlow in 1995, is in Charles G. Gross, "Genealogy of the 'Grandmother Cell'," *Neuroscientist* 8 (2002): 84–90.

15. Horace Barlow, "Single Units and Sensation: A Neuron Doctrine for Perceptual Psychology?," *Perception* 1 (1972): 371–394.

Up to now we have discussed how neurons represent what we see. We seem to have wandered away from our main topic—memory—but, as we argued in the previous chapter, perception and memory are intimately linked. The existence of "grandmother cells" would be fascinating, since it would imply that the memories I have of my mother, my grandmother, or a rose are stored in a single neuron or in a small aggregate of them. Since the discovery of face cells in the 1960s, many researchers have done experiments with monkeys, trying to find neurons whose responses are specific to a particular person or thing. (The failure of such attempts led to the favoring of a dense, holistic coding in which individual neurons encode minute details rather than abstract concepts and the representation of an object is given by the collective activation of a large number of neurons.) There is, however, a more interesting connection between perception and memory: the higher visual areas are connected directly to the medial temporal lobe and the hippocampus, which, as we saw when describing Patient H.M., is of critical importance for memory. Is it then the case that neurons in the hippocampus respond to visual stimuli? And if that is the case, how are these responses? Do they represent details or do they represent something more abstract—such as concepts—that we can process and store in memory?

11

THE JENNIFER ANISTON NEURON

From the outset I must make clear that the recordings of single neurons that I will describe here were not performed in monkeys, cats, or rats, like those from previous chapters, but in human beings. The activity of the human brain is generally studied using noninvasive, surgery-free techniques, such as electroencephalography or magnetic resonance imaging, which provide rather imprecise and indirect information about the firing of neurons. The monitoring of individual neurons, on the other hand, requires the insertion of electrodes in the brain, something routinely performed in animals but obviously not normally possible with human subjects. How, then, can single-cell studies be carried out in humans?

Between ten and twenty percent of epileptic patients suffer from seizures that cannot be controlled with medication. In some cases these seizures bring about a significant decline in the patient's quality of life, and if they happen to originate—i.e., have their focus—in nonvital parts of the brain, the possibility can be considered of removing the epileptic focus surgically. Patients that undergo this kind of surgery have a high probability of healing without any significant side effects. Before attempting the surgery, though, it is obviously necessary to locate the epileptic focus. In some cases this can be done based on clinical evidence and magnetic resonance imaging, but in others it is necessary to implant electrodes in the cranium to delimit the epileptic focus as accurately as possible. The decision of where to implant the electrodes is a strictly clinical matter, but often these are placed in the hippocampus and the surrounding structures—what is known as the medial temporal lobe—and once the electrodes are placed, we are fortuitously able to study the activity of this part of the brain in detail.

Technological innovations developed at UCLA (mostly having to do with the design of the electrodes) have resulted in recordings that allow us to see the activity of individual neurons in the human brain. The unique chance to perform such studies was one of the main things that enticed me to California ten years ago as a Sloan fellow. Though I certainly did not perform these studies on my own—some of the scientists with whom I collaborated include my mentor Christof Koch at Caltech and Itzhak Fried, one of the neurosurgeons who established this line of research at UCLA[1]—I will switch to the first person and describe without too many technical details my research on the neuronal activity in the human hippocampus.

Based on the connection between the hippocampus and the higher visual areas (and following earlier results obtained by Gabriel

1. Gabriel Kreiman, Leila Reddy, and Alexander Kraskov also participated in these experiments.

Kreiman, a friend and colleague from Argentina, now a professor at Harvard),[2] I started by looking for responses to visual stimuli, in particular those stimuli that were familiar or relevant to the patients. The reason is simple: in principle one expects that more neurons will respond to an image of the patient's mother than to a picture of someone they don't know. This initial intuition turned out to be correct. For example, in a patient who happened to be a soccer fan I found a neuron that responded to Argentine player Diego Armando Maradona (see figure 11.1); with a patient obsessed with the *Rocky* films I tried different characters from the series until I found a neuron that fired when we showed an image of Mr. T; in yet a third patient, one who loved to watch documentaries from the Discovery Channel, I found a response to pictures of animals.[3]

2. Gabriel mainly showed that in the medial temporal lobe there are responses to different kinds of visual stimuli (faces, objects, animals). Even more interesting is the fact that the neurons also fired when the patients closed their eyes and imagined the stimuli. These results are published in:
•   Gabriel Kreiman, Christof Koch, and Itzhak Fried, "Category-Specific Visual Responses of Single Neurons in the Human Medial Temporal Lobe," *Nature Neuroscience* 3 (2000): 946–953;
•   Gabriel Kreiman, Christof Koch, and Itzhak Fried, "Imagery Neurons in the Human Brain," *Nature* 408 (2000): 357–361.

3. The fact that there is a higher probability that neurons will respond to stimuli that are familiar and relevant to the patient is demonstrated in Indre Viskontas, Rodrigo Quian Quiroga, and Itzhak Fried, "Human Medial Temporal Lobe Neurons Respond Preferentially to Personally-Relevant Images," *Proceedings of the National Academy of Sciences (USA)* 106 (2009): 21329–21334.

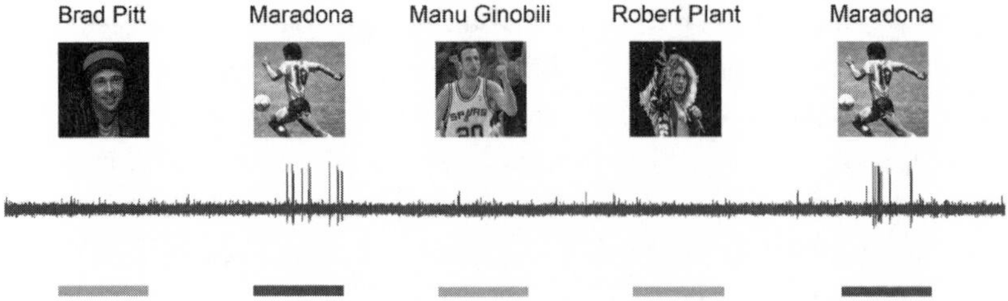

FIGURE 11.1
Recording from a single neuron that responded to Argentine soccer player Diego Armando Maradona but not to American actor Brad Pitt, Argentine basketball player Manu Ginóbili, British singer Robert Plant, or more than 50 other photos. The horizontal bars show the time (one second) that each photo was displayed on a laptop screen facing the patient.

These examples, and several others showing similar results, make it clear that neurons in the medial temporal lobe respond to visual stimuli. Now the question is whether the neuron in figure 11.1 responded to Maradona himself or to some detail in that particular photo—for example, the neuron might have responded to the green turf, a reasonable (though disappointing) possibility given the existence of neurons in monkeys' visual areas that respond to colors. The neuron might also have fired to the colors of the Argentine national team, the motion implied by Maradona's posture, or simply the ball. How can one prove that the neuron is responding to the person and not to some detail? Very simple: we show many pictures of Maradona (in different environments, with different background colors, in different postures, with different attire, etc.) and see if the neuron responds equally to all the photos (in other words, to the concept of

Maradona) or only to that particular picture. We found that it indeed responded to the concept.[4]

The first (and by far the most famous) of these neurons was one that responded to seven completely different images of actress Jennifer Aniston and to no other stimulus (including other people, animals, or places).[5] Figure 11.2 shows the neuron's response to 12 of the 87 photos shown in this particular experiment. The neuron fired when presented with different pictures of Aniston, but not when shown other celebrities like Kobe Bryant, Julia Roberts, Oprah Winfrey, or Pamela Anderson, places like the Golden Gate Bridge or the Eiffel Tower, or different animals. Another neuron from the same patient responded to different images of the Sydney Opera House, and yet another to the leaning tower of Pisa. The patient knew all of those people and landmarks quite well.

At this point it seems that the neurons in the hippocampus encode concepts such as a person or a particular place. To give even more conclusive evidence, we can take advantage of the fact that these experiments were performed in human subjects and see how the neurons respond to the names of the persons or objects eliciting responses. Figure 11.3 shows a neuron in the hippocampus of a patient that responded to different pictures of actress Halle Berry

4. More details can be found in Rodrigo Quian Quiroga, Alexander Kraskov, Christof Koch, and Itzhak Fried, "Explicit Encoding of Multimodal Percepts by Single Neurons in the Human Brain," *Current Biology* 19 (2009): 1308–1313.

5. Rodrigo Quian Quiroga, Leila Reddy, Gabriel Kreiman, Christof Koch, and Itzhak Fried, "Invariant Visual Representation by Single Neurons in the Human Brain," *Nature* 435 (2005): 1102–1107.

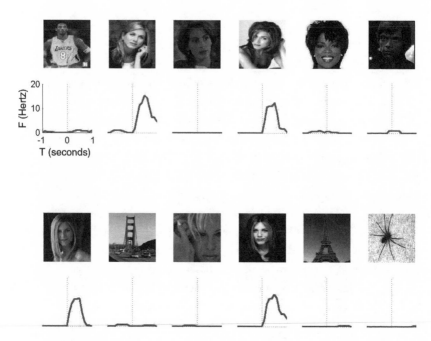

FIGURE 11.2

Responses of a neuron in the hippocampus that fired in response to different photos of Jennifer Aniston (only four out of seven pictures are shown here, but the responses to the other three were similar) and did not fire to images of other people, places, or animals (we show only eight of 80 pictures that were used). The thick lines show the average response to six presentations of each picture. The vertical axis shows the firing frequency in hertz (the number of spikes per second) and the horizontal axis shows the time (each image was shown for one second, starting at time zero).

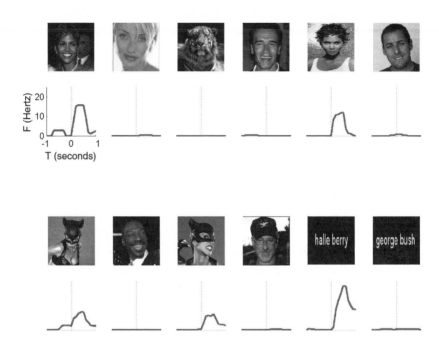

FIGURE 11.3

Responses of a neuron in the hippocampus to different pictures of Halle Berry, to Berry disguised as Catwoman, and to her name spelled out on the computer screen. The neuron did not respond to any other names or any other pictures of people, animals, or places.

and to her written name (but to no other names). At the time the experiment was performed, Halle Berry was promoting one of her films, *Catwoman*, and the neuron also fired when presented with pictures of her in costume, even though her face was not visible. In total, the neuron responded to four different pictures of Halle Berry, to three of her in costume, to a caricature of her, and to her name spelled out on the screen, but failed to respond to five caricatures of other people, seven other names, and 78 pictures of people (including Cameron Diaz, Arnold Schwarzenegger, and Adam Sandler), animals, and landmarks.

As we saw in the previous chapter, visual perception starts in the retina and goes through a neuronal circuit comprising the primary visual cortex and higher visual areas, ultimately reaching the hippocampus. The perception of auditory stimuli, on the other hand, follows a completely different route: it starts in cells located in the cochlea (in the inner ear) and goes through primary cortical areas in the temporal lobe before it also reaches the hippocampus. The question then arises: do the neurons we just described also respond to auditory stimuli? (And this is something we could test quite easily—all we have to do is say a name.) After all, the concepts we handle with our brains usually involve more than just one sensory modality. For example, if I see a photo of Einstein, if I see his name in the newspaper, or if I hear it on the radio, I will always grasp the same concept: Albert Einstein. Figure 11.4 shows the firing of a neuron that responded to three pictures of Luke Skywalker (from *Star Wars*) and to his name spelled out on the screen and spoken by a synthesized computer voice. The response shown in the figure was to a masculine voice, but a similar firing pattern was observed

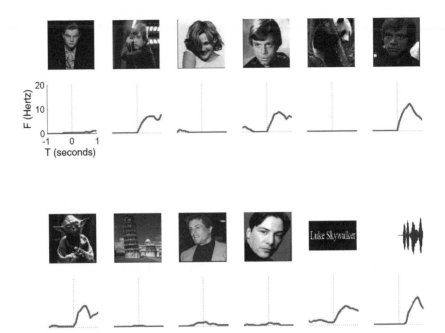

FIGURE 11.4

Responses of a neuron in the entorhinal cortex (an area of the medial temporal lobe that lies close to the hippocampus) to three pictures of Luke Skywalker and to his name, written and spoken (lower right panels). The neuron also responded to Yoda, another character from *Star Wars*.

when a feminine voice was used. This neuron did not respond to any other names, whether written or spoken, or to any other photos (except Yoda's, as we discuss below).

The responses in figure 11.4, as well as many other examples,[6] clearly show that the representation given by the neurons in the hippocampus (and the surrounding areas) is so abstract that it can be recalled with different types of stimuli. This implies that these neurons store and integrate information provided by the different senses in order to codify the semantic, abstract content of what we perceive. Interestingly, the neuron that fired in response to Luke Skywalker responded also to Yoda, another Jedi knight from the *Star Wars* saga. This is no coincidence; on the contrary, it is a common occurrence: if a neuron in one of these areas responds to more than one concept, these concepts will be related. For example, figure 11.5 depicts a neuron's responses to pictures of myself and to my name, both spelled out on the screen and spoken out loud by the computer, as well as to the photos and names of three colleagues who also performed experiments with this patient at UCLA.

Beyond giving another clear example of the responses of these neurons to visual stimuli, and showing again how they fire to associated concepts and even encode conceptual categories (the four researchers at UCLA), the neuron of figure 11.5 provides further crucial evidence on the way the hippocampus functions. The key fact here is that the patient did not know us (neither me nor my

6. More examples appear in Rodrigo Quian Quiroga, Alexander Kraskov, Christof Koch, and Itzhak Fried, "Explicit Encoding of Multimodal Percepts by Single Neurons in the Human Brain," *Current Biology* 19 (2009): 1308–1313.

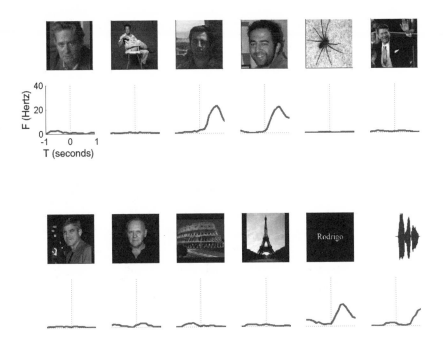

FIGURE 11.5

Responses of a neuron in the hippocampus to different pictures of myself and to my name, Rodrigo, both spelled out and spoken by the computer. This neuron had similar responses when presented with the photographs and names of three colleagues who performed studies with this particular patient.

colleagues) and obviously did not know our names until a couple of days before the study was carried out. This implies that such an abstract representation may be generated relatively quickly, within a day or two, probably in minutes or even seconds. We are not born with neuronal responses like those shown in the previous figures; we create them as needed, as we process different concepts at a given time. And, in the same way that responses in the hippocampus are generated relatively fast, they can also be lost in time if they become irrelevant. For example, I doubt that the patients I saw years ago have many neurons left that would respond to an image of me or of any of my colleagues at UCLA. Maintaining them would, in fact, be a waste of resources, since it is not likely that these patients will ever see us again. It would be more useful if these neurons were instead available to encode something important taking place in the patients' present situation. Here we see once more the importance of forgetting.

I have already mentioned Wim Wenders's film *Until the End of the World*, in which a scientist attempts to embed images in his blind wife's brain and, once he achieves this, seeks to invert the process, interpreting her brain signals in order to project thoughts onto a screen. This was one of my favorite movies when I was a physics major: it had a terrific soundtrack, a very intriguing idea impeccably explored, and an avant-garde aesthetic that later reappeared in a series of U2 videos also directed by Wenders. Almost twenty years after I saw the film for the first time, fate put me in the position of doing something similar but for real: based on the neuronal responses to Jennifer Aniston, Halle Berry, and the like, I wondered if I could predict what the patient was seeing each time. Leaving technical details aside, the idea is quite simple: if I see that the Jennifer Aniston neuron fires, I predict the patient is seeing a photo of Jennifer Aniston; if instead the Halle Berry neuron fires, I conclude that he is seeing Halle Berry, and so on. In truth it was necessary to monitor the activity of more than a couple of neurons, and instead of making predictions by eye I had to resort to decoding algorithms from the field of machine learning. The result was surprising: the firing of a relatively small number of neurons sufficed to predict (much better than by random guessing) which photo the patient was seeing.[7] But this is not the end of the story. Using the same principle, in an experiment led by Moran Cerf, a student of Christof Koch's at Caltech,[8] we showed that patients were able to project their thoughts onto a display by voluntarily controlling the firing of their neurons. We did so with a decoding algorithm that interpreted the neurons' activity. Starting with hybrid images—superposing semitransparent photos of, say, Jennifer Aniston and Diego Maradona—the patient was able, by thought alone, to bring into focus one picture at the expense of the other. This takes me back to a scene from *Until the End of the World* in which Henry Farber, the scientist played by Max von Sydow, and Claire (played by Solveig Dommartin) can barely contain their ecstasy as they see images from her brain trickling onto the screen.

7. Rodrigo Quian Quiroga, Leila Reddy, Christof Koch, and Itzhak Fried, "Decoding Visual Inputs from Multiple Neurons in the Human Temporal Lobe," *Journal of Neurophysiology* 98 (2007): 1997–2007.

8. Moran Cerf, Nikhil Thiruvengadam, Florian Mormann, Alexander Kraskov, Rodrigo Quian Quiroga, Christof Koch, and Itzhak Fried, "Online, Voluntary Control of Human Temporal Lobe Neurons," *Nature* 467 (2010): 1104–1108.

I should clarify something: Our methods only project what the patient voluntarily wants to show us (based on photos previously found to elicit a response in a neuron), something that requires effort and focus. We made no attempt to pick their brains; we just developed an alternative mode of communication: instead of saying "Jennifer Aniston," the subject could directly project her image just by thinking of her. Why is this important? First, because it shows that a person has the ability to control at will the activity of individual neurons in their brain, simply by thinking. (This makes us reject Cartesian dualism more than ever: mind and neurons are one and the same thing.) Second, this technology could be used in the future in patients with locked-in syndrome or other severe motor disorders—like Stephen Hawking, the famous physicist with Lou Gehrig's disease—who have a limited ability to move and hence to communicate, either by talking, writing, or even moving a finger. In a few years, then, it is possible that these patients will be able to communicate directly with the outside world by using their neurons' activity. Similarly, patients with paralysis or with amputated limbs could use the activity of their neurons to control, say, a prosthetic arm.[9]

9. This line of research has been explored by Richard Andersen, who, along with Christof Koch, was my mentor at Caltech. More details can be found in:
• Richard A. Andersen, Joel W. Burdick, Sam Musallam, Bijan Pesaran, and Jorge G. Cham, "Cognitive Neural Prosthetics," *Trends in Cognitive Sciences* 8 (2004): 486–493;
• Mikhail A. Lebedev and Miguel A. L. Nicolelis, "Brain-Machine Interfaces: Past, Present and Future," *Trends in Neuroscience* 29 (2006): 536–546.

In the previous chapter we discussed different types of neuronal coding schemes. As we saw, a given concept might be represented by the firing of either many neurons (a dense, holistic representation), relatively few neurons (a sparse representation), or a single neuron (a straw man version of the "grandmother cell" coding). The neurons in the hippocampus clearly do not use a dense representation—in which each neuron encodes different details: Jennifer Aniston's hair color, her eye color, the width of her lips—but rather a more abstract one. Are these, then, the long-sought grandmother cells? Let's see . . . If there was one and only one neuron responding to Jennifer Aniston, the probability of finding it among billions of other neurons would be close to nil. The simple fact that we found a neuron that fired to Jennifer Aniston implies, by plain and simple statistics, that there must be more. After all, in the famous parable that gave grandmother cells their name, Lettvin never said that there was a single neuron for each concept—in fact he suggested there were 18,000. Is it then possible that there are many neurons—thousands, perhaps hundreds of thousands—that respond to one and only one person? It could be, but if I find a neuron that responds solely to Jennifer Aniston, I cannot rule out that this same neuron may have fired in response to some other person that was not shown during the experiment. In fact, the Jennifer Aniston neuron also responded to Lisa Kudrow, another star of *Friends* (the television series that made them both famous), in an experiment carried out the next day. Moreover, we have already seen that the Luke Skywalker neuron responded also to Yoda and that the neuron that responded to photos of me did the same when shown pictures of my UCLA colleagues. One could argue that these neurons are responding to somewhat more elaborate concepts (the blondes in

Friends, the Jedi knights in *Star Wars*, the researchers at UCLA), which would make them in effect grandmother cells. Whether or not to call them grandmother cells is in the end a semantic question that, in my view, has little scientific interest. In practice I prefer to avoid the name because it usually gives rise to confusing and incorrect interpretations.[10]

In summary, we cannot say that there is one and only one neuron responding to a given concept, and neither can we say that there is a group of neurons responding to one and only one thing. We can, nonetheless, state that in the areas that we study there are groups of relatively few neurons (though not just one) that respond to a small number of concepts. In other words, we have a sparse representation or, as Horace Barlow called them, "cardinal neurons." Although it may be disappointing not to call them "grandmother cells,"[11] we can still note two features that make these neurons very interesting: first, they respond to abstract concepts, not to details found on a particular picture; second, if they respond to more than

10. See, for example, the discussions that I published with Gabriel Kreiman:
• Rodrigo Quian Quiroga and Gabriel Kreiman, "Measuring Sparseness in the Brain: Comment on Bowers (2009)," *Psychological Review* 117 (2010): 291–297;
• Rodrigo Quian Quiroga and Gabriel Kreiman, "Postscript: About Grandmother Cells and Jennifer Aniston Neurons," *Psychological Review* 117 (2010): 297–299.

11. It is, though, quite tempting to claim to have discovered grandmother neurons, given the discussion they have generated among neuroscientists and the many attempts to find them.

one concept, these concepts are related (Jennifer Aniston and Lisa Kudrow, Luke Skywalker and Yoda, etc.).[12]

Is there then a sparse representation of concepts throughout the brain? No, it would be a mistake to state this in general. As we saw in the previous chapter, the neurons in the retina respond to the intensity of light at a given point, while those in the primary visual cortex react to lines with a given orientation. None of these neurons can by itself tell what image we are seeing; the representation of the object is formed by the collective activity of a large number of them. This is a dense, holistic, and implicit representation. The higher visual areas integrate the information given by these neurons and transmit it to the hippocampus, whose neurons respond to concepts by means of a sparse and explicit representation (i.e., a single neuron can tell that a give concept is being seen).

Now, why does the brain go from a dense, implicit, and detail-oriented coding at the visual cortex to a sparse, explicit, and concept-based coding at the hippocampus? To understand this we must recall our discussion of Patient H.M., whose hippocampi were surgically removed to cure the uncontrollable epilepsy that ailed him. H.M. was afterward incapable of forming new memories but was still able to identify different people and objects; this means

12. More details can be found in:
• Rodrigo Quian Quiroga, Gabriel Kreiman, Christof Koch, and Itzhak Fried, "Sparse but Not 'Grandmother-Cell' Coding in the Medial Temporal Lobe," *Trends in Cognitive Sciences* 12 (2008): 87–91;
• Rodrigo Quian Quiroga, Alexander Kraskov, Christof Koch, and Itzhak Fried, "Explicit Encoding of Multimodal Percepts by Single Neurons in the Human Brain," *Current Biology* 19 (2009): 1308–1313.

FIGURE 11.6
The brain goes from a dense representation of points and contours in the retina and the primary visual cortex (top panels), to a more complex coding in the higher visual areas (bottom left), and then to a sparse representation of the concept, the identity of the person, in the hippocampus (bottom right).

that the hippocampus is not necessary for visual recognition. We also know that the hippocampus is not the final (and unique) area of memory storage, since H.M. had memories from the time preceding his surgery. What, then, does the hippocampus do? What role do neurons like the Jennifer Aniston neuron play? Neither perception nor (factual) memory storage—the hippocampus makes the connection between the two. I do not need the Jennifer Aniston neuron to recognize her picture if I see it in a magazine—this is done by the neurons in the inferior temporal cortex—but I do need it to transform that perception into a memory, so that in the future I can remember having seen that picture of Jennifer Aniston in the magazine. Neurons like the one for Jennifer Aniston help us abstract concepts that we can use to create new links and memories. The conceptual representation by these neurons makes much sense since in general we tend to remember generalities about people, deeds, and places, and forget most of the surrounding minutiae. In the last few pages I have strayed far from Borges, but at this point we come full circle. If we did not have this type of neurons we would end up like Funes the memorious, without the capacity to abstract or even to think, remembering only irrelevant details.

The fact that the neurons in the hippocampus respond to associated concepts is perfectly consistent with this interpretation. After all, we tend to make associations as we form new memories and uncover new relations. We associate Luke Skywalker with Yoda, since they both are *Star Wars* characters. If a friend introduces me to his girlfriend—let's call her Clara—I will associate the two of them through the firing of neurons in my hippocampus. This is also the basis for episodic memories (i.e., memories of events) discussed in the previous chapters. If I met my friend and his girlfriend at the theater watching *Until the End of the World*, then some of the neurons that respond to them will also respond to the film so that I can associate those concepts and remember the occasion. To think is to abstract, to associate concepts, to create relations and categories. These associations allow us to consolidate memories and keep them alive. Perhaps months later I may have forgotten the name of my friend's girlfriend, but once I remember I met them at the theater and recall that we watched *Until the End of the World*, I may well remember that her name was similar to Claire's from the film. These ideas are not new: already in the late nineteenth century William James argued that associations strengthen memories:

"If we have not the idea itself, we have certain ideas connected with it. We run over those ideas, one after another, in hopes that some one of them will suggest the idea we are in quest of; and if any one of them does, it is always one so connected with it, as to call it up in the way of association." . . . The 'secret of a good memory' is thus the secret of forming diverse and multiple associations with every fact we care to retain.[13]

A clear example of how associations help bring up memories takes place when we wake up: if we remember something that happened in a dream, we can associate it to something else and this to yet something else and so on, to remember at least part of what we have dreamed. On the other hand, if we do not have a detail with which to start weaving the tangle of associations, our memory goes blank. This wandering from one concept to the next through associations—this flow of consciousness—is what led Proust to reconstruct his complete boyhood after tasting a madeleine. In the same way that the neurons described above create connections between concepts, they can also create this flow from one concept to another. Jennifer Aniston neurons are then the building blocks of explicit memory functions. Going back to the encounter at the movies, I may start by remembering my friend, then remember I watched a

13. William James, *The Principles of Psychology*, authorized ed., vol. 1 (New York: Henry Holt, 1890; repr., New York: Dover, 1950), pp. 653 and 662. In the first half of this passage James is quoting from James Mill, *Analysis of the Phenomena of the Human Mind*, vol. 1 (London: Baldwin and Craddock, 1822), p. 235.

film with him, then recall that his girlfriend had a similar name to one of the main characters, and so on. In light of this interpretation, it is not surprising that the neurons in the hippocampus (or, in general, in the medial temporal lobe) respond only to concepts that are important. They simply act in response to something that is familiar, something relevant enough to be stored in memory. After all, we do not want to remember everything; we do not want to end up like Funes.

KEYS TO THOUGHT

Let us start this final chapter by discussing chess, a topic to which Borges dedicated two sonnets published in *The Maker* in 1960. The first of them says:

En su grave rincón, los jugadores
rigen las lentas piezas. El tablero
los demora hasta el alba en su severo
ámbito en que se odian dos colores.

Adentro irradian mágicos rigores
las formas: torre homérica, ligero
caballo, armada reina, rey postrero,
oblicuo alfil y peones agresores.

Cuando los jugadores se hayan ido,
cuando el tiempo los haya consumido,
ciertamente no habrá cesado el rito.

En el Oriente se encendió esta guerra
cuyo anfiteatro es hoy toda la tierra.
Como el otro, este juego es infinito.[1]

[In their grave corner, the players
rule the slowly moving pieces. The board
keeps them till dawn in its ruthless confines
in which two colors hate each other.

Inside the game the forms shine
their magical essence: the Homeric rook,
the agile knight, the armored queen, the solitary king,
the oblique bishop, the attacking pawns.

After the players are gone,
after time has made them vanish,
the rite will certainly not be over.

In the East started this war
whose theater is now the whole earth.
Like the other, this game has no end.]

Figure 12.1 shows two chessboards with the same pieces but in different distributions. To someone who has not learned to play chess, there is little difference between the two. To someone who knows the game, though, the board on the right-hand side is much harder to memorize, since the pieces are laid out at random, while on the board at left they are distributed in a configuration that is logical after a few moves. (The knowledgeable reader will remember the configuration even more easily, as he is likely to identify the

1. Jorge Luis Borges, "Ajedrez [Chess]," in *Obras completas* (Buenos Aires: Emecé, 2007), vol. 2, pp. 226–227.

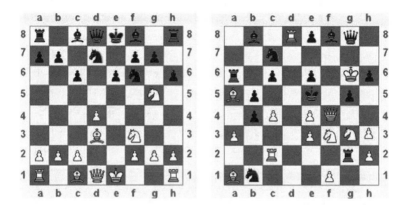

FIGURE 12.1

Thirty pieces distributed on a chessboard. The left panel shows a configuration that may occur in an actual game, while the right panel depicts a random placement. Chess players find it much easier to remember the distribution from the real game.

Steinitz variation—ECO B17—of the Caro-Kann Defense, with the typical advance of the black pawn in file c and the exchange of central pawns in files d and e.) This was verified by means of a famous experiment in which players of varying levels of expertise were shown a valid chess configuration for five seconds. The position had 26 of the 32 pieces in play, and while the experts were able to remember the locations of 16 pieces on average, the beginners managed to remember only four. The interesting finding, however, was that the level of expertise ceased to matter when the players were shown random configurations.[2] This result shows that we

2. William Chase and Herbert Simon, "Perception in Chess," *Cognitive Psychology* 4 (1973): 55–81.

tend to remember the meaning of things, their general and abstract structure. During a game, chess masters do not pay attention to the position of every single piece (as a beginner is forced to do) but to their general distribution on the board, the defense by the black pieces and the salient features of the white attack. Starting from a given configuration, a master will extract a meaning, he will see a Sicilian Defense or a Queen's Gambit, and from this generic representation he will plan a general strategy to follow as the game unfolds. He will try to break through the king's defense or take over the center of the board, but in neither case will he compute every last movement of the bishops, knights, or rooks.

The configuration in the left-hand panel of figure 12.1 may be one of the most famous in the history of chess; it shows the position right after the black pawn's advance in file h, Garry Kasparov's fatal mistake that cost him the sixth and last game against IBM's super-computer Deep Blue (Deep Blue played white, Kasparov black).[3]

---

3. White's advantage is far from obvious. In the next move, Deep Blue captured the pawn at e6 with the knight, sacrificing a piece but gaining a complex positional advantage. It is worth noting that in principle it is very unlikely that a computer will sacrifice a piece unless there is an obvious advantage to doing so, which was not the case here. However, Deep Blue also had in memory a deep cache of matches taken from books that led it to choose this move. The sacrifice of the knight was a hard blow to Kasparov, who immediately started to look resigned to his fate. His next move—advancing the queen to e7 instead of capturing the knight directly—was even less fortunate and clinched his defeat. Kasparov abandoned the game after 19 moves and slightly over an hour's play—not only his fastest defeat ever, but also the first time in his career that he gave up a match.

The significance of this match reached well beyond the world of chess. Kasparov, seen by many as one of the foremost examples of human intelligence, was defeated by a machine. Thus the question arose: can a machine capable of evaluating 200 million moves per second and anticipating any possible outcome 20 moves ahead be more than a human being?

What led Kasparov to his defeat is in fact one of the fundamental differences between the human mind and the computer. Starting perhaps with the second game (which he lost when he should have come to a draw), Kasparov was psychologically broken. He had lost his usual confidence and just wanted to finish each game as quickly as possible. This exhaustion was not physical—the games were fast and painless compared to his epic battles against Karpov in the eighties—but mental, and due to very complex reasons that even he could not completely fathom. This weakness, this "defect," is what makes us human, imperfect, unpredictable and fascinating. It is this aspect of human behavior that machines have not (yet) been able to reproduce because it cannot be reduced to logical rules and turned into an algorithm.

A chess grandmaster needs an amazing memory to be able to remember an enormous number of openings, variations, and previous games, and a remarkable computational power to be able to anticipate the outcome of a given move. But is this all? Can anyone with good memory and extraordinary computational ability become a chess grandmaster? As I sit here writing this sentence, I see that a portable memory chip that fits 12 gigabytes of information in a couple of square centimeters costs less than $30. Someone once estimated that a 30-year-old adult stores around 150 megabytes of memory,[4] about a hundredth of what the chip can store. Is then this chip, about the size of my thumbnail, more powerful than the human brain? Clearly not; we are much more than a collection of memories. Funes the memorious could probably beat more than one player simply playing "by the book." A chess program running on a supercomputer like Deep Blue got Kasparov himself into trouble with its impressive memory and processing power. Is this all, though? Certainly not, since the computer, no matter how sophisticated, uses brute force to decide its moves. It can follow any possible course of action 20 moves ahead, compute every single ensuing ramification, and choose the best one according to an informed set of rules, but it cannot think; it lacks the main quality that distinguishes the extraordinary player from the merely good: intuition. Kasparov, for example, was practically unbeatable in blitz chess.[5] In these games each move has to be made in a fraction of a second and players have no time to consider future configurations. They must rely almost entirely on their intuition and experience.

4. In a series of experiments resembling those carried out by Gustav Spiller and Frederic Bartlett (whom we shall discuss later on), Thomas Landauer tested the human ability to memorize pieces of text, photos, words, bits of music, etc., and found that on average a person remembers two bits per second. At this rate, the memory of a 30-year-old person who has been awake for 15 hours a day would sum up to 148 megabytes. More details can be found in Thomas Landauer, "How Much Do People Remember? Some Estimates of the Quantity of Learned Information in Long-Term Memory," *Cognitive Science* 10 (1986): 477–493.

5. In these matches each player has only a few minutes—between three and five, according to the category of play—to make all their moves.

They cannot explain why they made a particular move; they just have the gut feeling that it is the right one (and usually it is). Developing this intuition takes years of training, of making abstractions and developing unconscious strategies. Chess players do not see isolated pieces but an ensemble, a whole, a structure that can be stronger or weaker, solid in defense or deployed for attack, leading or lagging with respect to the development of the opponent's pieces.

In his "Argumentum ornithologicum," Borges says:

Cierro los ojos y veo una bandada de pájaros. La visión dura un segundo o acaso menos; no sé cuántos pájaros vi. ¿Era definido o indefinido su número? [. . .] En tal caso, vi menos de diez pájaros (digamos) y más de uno, pero no vi nueve, ocho, siete, seis, cinco, cuatro, tres o dos. Vi un número entre diez y uno, que no es nueve, ocho, siete, seis, cinco, etcétera.[6]

[I close my eyes and see a flock of birds. The vision lasts for a second or maybe less; I cannot say how many birds I saw. Was their number definite, or indefinite? . . . In any case, I saw fewer than (say) ten birds and more than one, but I did not see nine, eight, seven, six, five, four, three, or two. I saw a number between ten and one which is not nine, eight, seven, six, five, etc.]

These lines artfully summarize the point. To illustrate this idea consider figure 12.2. Give it a fleeting glance and look away before taking in any detail. How many birds are there? Are there five, nine, twelve, twenty, fifty? We cannot say with certainty after a quick look. We know we saw fewer than fifty birds, perhaps fewer than thirty or even twenty, and we also know we saw more than five, perhaps more than ten birds. As Borges would say, we saw a number

6. Jorge Luis Borges, "Argumentum Ornithologicum," in *Obras completas*, vol. 2, p. 198.

FIGURE 12.2
Looking fast and without counting, ask yourself: How many birds are there in the photograph?

between ten and twenty that is not ten, eleven, twelve, thirteen, etc. It is really surprising that we have an idea of how many birds there are in the figure even though we were not able to count them. This is what makes us human, what no machine can copy: our capacity to intuit, to have impulses and gut feelings (which in fact turn out to play a key role in many of the decisions we make throughout our lives). A computer can easily count the number of birds in the picture—perhaps faster than the duration of our blinking glimpse —but it cannot estimate the number based on intuition. We acquire this intuition based on our unique capacity for abstraction and generalization after seeing sets of a similar size countless times, be they of oranges, cars, or wooden matches.

The examples of the chess game and the flock of birds illustrate two fundamental characteristics of our thought and memory that we have often discussed in this book. The first is the importance of abstraction, of generating meaning and concepts. The second is the importance of forgetting. These two principles are intimately linked, since the very act of abstracting implies neglecting and forgetting details.

As we saw in the previous chapter, this abstraction, this loss of detail, reaches its pinnacle in the hippocampus. The Jennifer Aniston neuron responded equally to completely different images of the actress: it did not matter if she was shown from the front or in profile, if she looked thoughtful or was smiling, if she was bare-shouldered or wearing a black coat. Furthermore, neurons in the hippocampus also fired to the written or spoken name of the actress. The parallel with "Funes the Memorious," the story that Borges imagined more than 50 years before these studies, is truly fascinating. We can see Funes as someone lacking those "Jennifer Aniston neurons" that encode abstract concepts. Writes Borges of Funes:

No sólo le costaba comprender que el símbolo genérico perro abarcara tantos individuos dispares de diversos tamaños y diversa forma; le moles-taba que el perro de las tres y catorce (visto de perfil) tuviera el mismo nombre que el perro de las tres y cuarto (visto de frente).[7]

[Not only was it difficult for him to understand that the generic term "dog" could embrace so many disparate individuals of diverse size and shapes; it bothered him that the dog seen in profile at 3:14 would be called the same as the dog at 3:15 seen from the front.]

7. Jorge Luis Borges, "Funes el memorioso," in *Obras completas*, vol. 1, pp. 583–590.

We have discussed Patient H.M. and seen how his unfortunate case gave us solid evidence of the role played by the hippocampus in the formation of memories. Thus the abstraction shown by neurons like the one for Jennifer Aniston in the hippocampus serves to encode concepts in the abstract form in which they will be used by memory functions. Such a representation is indeed ideal for associating concepts, for generating relationships, for thinking.[8]

In March 2010, along with María Kodama I took part in a symposium about Borges and memory organized by the Physics Department at the University of Buenos Aires. The talks were delivered in the main auditorium of Pavilion I on campus, a lecture hall in which I had sat through at least three courses and had taken many exams. Before my talk I tried to remember, twenty years later, some of the many experiences I had lived through during the hundreds of hours that I spent in that room. But I barely managed to remember a couple of things that I can recount in under a minute: I remembered Adrián Paenza, back then a linear algebra instructor, vividly illustrating a transformation between Cartesian coordinate frames by moving the origin from his navel to his head; I also remembered a comment made by an algebra lecturer as he was explaining the sieve of Eratosthenes (a simple rule to generate the prime numbers): he joked that "the sieve of Eratosthenes" was a

8. Given an abstract representation, associating two concepts requires establishing connections only between the neurons that encode the concepts. If neurons encoded details, then associating the concepts would require establishing connections between two sets of details, a much more complex task.

trivial procedure with a pompous name, and that claiming acquaintance with it was a cheap way to earn fame as a genius.

This simple anecdote—which in fact I used at the beginning of my talk—illustrates several interesting aspects of what we have been discussing. First of all, we tend to forget an enormous amount of detail. After spending hundreds of hours in that room I can recall only a couple of events that took place in it, and it takes me mere seconds to reconstruct them. As we saw in chapter 2, Gustav Spiller observed this more than a century ago as he attempted the monumental task of enumerating everything he remembered (something of which Borges took note). A similar argument was later developed by Frederic Bartlett (1886–1969) at the University of Cambridge. Bartlett asked his students to memorize different texts and observed that their recollections tended to be shorter than the originals and, moreover, were molded to suit the subjects' personal interpretations.[9] Second, the few memories I have of the hours I spent in the auditorium are very abstract and shorn of any detail. I do not remember what Paenza was wearing or how he looked like when he gave that memorable lecture; I only remember concepts that I associate in my memory: Paenza, the auditorium, the lecture on coordinate transformations. Of the other instructor I cannot even recall the name; in my memory he remains "the Algebra I instructor," the funny guy that once made the joke about the sieve of Eratosthenes. Third, the little I remember is those events which for some reason ended up being salient and relevant enough to be etched in my memory. In this regard, we have already seen that

9. Frederic Bartlett, *Remembering* (Cambridge: Cambridge University Press, 1932).

neurons in the hippocampus tend to fire in response to people or things that are familiar and important to the person (Mr. T for the *Rocky* fan or Maradona for the fan of soccer).

The process of abstraction we are referring to already starts with visual perception—but the same applies to the other senses as well—with the creation of signs described by Helmholtz in the nineteenth century. We do not process images in our brains in the same way a camera does; on the contrary, we extract a meaning and leave aside a multitude of details. This gives rise to phenomena like change blindness or perception deficits, as in the "spot the differences" game (see chapter 9). Moreover, the sense we make of what we perceive depends on our experience and on a large number of unconscious inferences. When we look at the picture with the birds we do not even count them; we see them as a flock of some ten or twenty birds. We do not know exactly how many birds there are or remember the position of each of them because we do not care; we do not want to process, think about, or store such information. This process of abstraction that starts with perception then carries forward to memory. Bartlett noticed it clearly when he saw that his students did not remember literally the stories that he asked them to read, but rather remembered a *scheme*, the meaning they extracted from the story based on their personal experience. Just like perception, memory is a creative process; when we remember something we do not repeat the experience as it was; rather, we relive it in another context, create a new representation, and even change its meaning (as Elizabeth Loftus showed with the different recalls that subjects gave of the scene of a traffic accident; see chapter 8).

Perception and memory are inextricably linked. We cannot recognize an object if we do not remember it from before. The brain must store at least some details somewhere if we are to recognize a person, a tree, or a cup of tea; but we certainly do not store everything—we extract particular features that help us recognize these people and objects in very different circumstances. Recognizing someone is only the beginning, the trigger that sets into motion a cascade of processes that involve memories and emotions. Starting with perception, the process of thinking requires making abstractions; otherwise we would end up like Funes, crammed full of details but incapable of making even minimal associations. This saturation with detail leads to the "verbal adhesion" described by Langdon Down, the ability to memorize something but without understanding its content. This is one of the main characteristics of autistics and savants. It is what bedeviled Shereshevskii, who could repeat a long list of objects without any mistakes but was unable to say which ones were liquids. The ability to abstract and generalize is what distinguishes Enzo Gentile, the mathematician mentioned in chapter 7, who could explain like no one else the underpinnings of Fermat's last theorem but fumbled when attempting a simple subtraction, from a savant with an extraordinary ability for calculation like Daniel Tammet, who could memorize 20,000 digits of π or compute the fourth power of 37 in mere seconds but is at a loss when trying to understand an equation.[10]

The importance of abstractions was already established by Plato when he postulated the existence of "universals"—abstract ideas that go beyond the particular instances that we perceive with our senses—with his famous allegory of the cave. Aristotle refuted his

---

10. Daniel Tammet, *Born on a Blue Day* (London: Hodder and Stoughton, 2006), p. 117.

master when he conceived universals not as transcendent entities but as abstractions that we make of things, *images* we create in our minds starting from the experience of what we perceive with our senses. Aristotle stated in *On the Soul* (431a, 431b) that knowing something means creating a concept, extracting a set of characteristics of an object in order to abstract its essence. These processes of abstraction and generalization are what allow us to form the ideas on which our thought is based.

> To the thinking soul images serve as if they were contents of perception (and when it asserts or denies them to be good or bad it avoids or pursues them). That is why the soul never thinks without an image. . . . And as in the former case what is to be pursued or avoided is marked out for it, so where there is no sensation and it is engaged upon the images [the thinking soul] is moved to pursuit or avoidance. E.g. perceiving by sense that the beacon is fire, it recognizes in virtue of the general faculty of sense that it signifies an enemy, because it sees it moving.[11]

Thomas Aquinas later took up Aristotle's ideas and described in more detail how the meaning that we give to things, the *images* or *phantoms* we create in our minds, are formed from different abstractions. In a monumental effort to make Aristotle's philosophy compatible with the doctrine of the Church, Aquinas stressed the distinction (already made by Aristotle) between passive intellect, the one that allows us to perceive and that we share with animals, and active intellect, exclusive to human beings, that allows us to think and understand what we perceive.

11. Aristotle, *De anima* (*On the Soul*), trans. J. A. Smith, in *The Basic Works of Aristotle*, ed. Richard McKeon (New York: Random House, 1941), p. 594.

The process of abstraction starts with perception itself, with the extraction of signs described by Helmholtz. From those signs we generate concepts and memories, the schemes described by Bartlett. Neither Helmholtz's signs nor Bartlett's schemes are reproductions of visual images or the exact reconstructions of a chain of events forming a memory. On the contrary, both are constructions that we elaborate starting from the meaning that we attribute to things based on our personal experience. This construction is based on how we define and categorize concepts. If I see a dog in the park, I may think, "There goes a dog." Someone else with a little more knowledge may think, "There goes a Labrador," while a third person may say, "There goes my dog." Thus one and the same dog, the same stimulus in three different people, generates three different signs: a dog, a Labrador, my dog. The first refers to a type of animal, the second to a much more specific group, and the third identifies an individual. Furthermore, given our differences in perception and our diverse experiences, the same dog playing in the park will generate completely different memories in the three of us.

Different levels of categorization give rise to different interpretations, and the meaning we give to things is precisely what the neurons in the hippocampus represent. In the previous chapter we saw that a neuron fired in response to images of four researchers (myself included) who performed experiments at UCLA. This neuron encodes a category involving different subjects ("the researchers"). At the same time, there certainly must have been other neurons that encoded each of us separately so the subject could have more specific memories, like "Rodrigo is the one with the Spanish accent." The variety of meanings furnished by different levels of categorization is a subject explored by Borges in "Keats's Nightingale" (from

1952's *Other Inquisitions*). In his "Ode to a Nightingale," John Keats writes, "Thou wast not born for death, immortal Bird! / No hungry generations tread thee down; / The voice I hear this passing night was heard / In ancient days by emperor and clown." Quoting Sydney Colvin, Borges argues that there is a lapse of logic in these lines because Keats refers to the Platonic nightingale, the species, whose immortality contrasts with the transience of human life—not the life of the species but the life of each individual.

Categorization—the extent to which we isolate a meaning and discard a myriad of details—manifests itself in language through the use of nouns. In a remarkable passage ("Blather for Verses," from *The Size of My Hope*, 1926), Borges says:

El mundo aparencial es un tropel de sensaciones barajadas. . . . El lenguaje es un ordenamiento eficaz de esa enigmática abundancia del mundo. Dicho sea con otras palabras: los sustantivos se los inventamos a la realidad. Palpamos un redondel, vemos un montoncito de luz color de madrugada, un cosquilleo nos alegra la boca, y mentimos que esas tres cosas heterogéneas son una sola y que se llama naranja. La luna misma es una ficción. Fuera de conveniencias astronómicas que no deben atarearnos aquí, no hay semejanza alguna entre el redondel amarillo que ahora está alzándose con claridad sobre el paredón de Recoleta, y la tajadita rosada que vi en el cielo de la Plaza de Mayo, hace muchas noches. Todo sustantivo es una abreviatura. En lugar de contar frío, filoso, hiriente, inquebrantable, brillador, puntiagudo, enunciamos puñal; en sustitución de alejamiento del sol y profesión de sombra, decimos atardecer.[12]

12. Jorge Luis Borges, *El tamaño de mi esperanza* (1926; repr. Madrid: Alianza editorial, 2008), pp. 50–55. This passage is quoted by Eduardo Mizraji in "Memoria y pensamiento," included in *Borges y la ciencia* (Buenos Aires: Eudeba, 1999).

[The world of appearance is a jumble of shuffled sensations. . . . Language is an effective ordering of the world's enigmatic abundance. In other words, we attribute nouns to reality. We touch a round shape, we see a little lump of light the color of dawn, a tingling elates our mouth, and we lie to ourselves and say that these three disparate things are but one and that it is called an orange. The moon itself is a fiction. Apart from astronomical facts, upon which we will not dwell here, there is no resemblance whatsoever between the yellow circle now clearly rising above the Recoleta and the thin pink sliver that I saw above the Plaza de Mayo a few nights ago. Every noun is an abbreviation. Instead of enumerating cold, sharp, hurtful, unbreakable, shiny, pointy, we say dagger; instead of the sun receding and the shadows approaching, we say dusk.]

In "Tlön, Uqbar, Orbis Tertius" (from *Ficciones*, 1944), Borges conceived a congenitally idealistic world whose language has no nouns (to idealists like Locke and Berkeley, abstractions of Platonic concepts are metaphysical aberrations, simple linguistic expressions). In the languages spoken in the southern hemisphere of Tlön, "'The moon rose over the river' translates as 'hlör u fang axaxaxas mlö,' or 'Upward behind ever-flowing mooned,' in that order." In the languages spoken in the northern hemisphere, nouns are formed by an accumulation of adjectives: "Instead of 'moon' one says 'light-airy-over-dark-round or tenuous-reddish-yellow of the sky.'"[13] Further on, Borges writes that such "total idealism" invalidates science, since in Tlön "every mental state is irreducible;

13. During a conference at Indiana University in 1976 someone observed that the original Spanish includes the word *cielo* ("sky"), which is in fact a noun, and that instead Borges could have used *celestial*. *Jorge Luis Borges: Conversations*, ed. Richard Burgin (Jackson: University Press of Mississippi, 1998), pp. 170–171.

the mere act of naming it, classifying it, implies a falsification. One could deduce from this that in Tlön there is no science, not even rational thinking."

The importance of categorization, of the use of general names, was recognized by John Stuart Mill in the nineteenth century (in a passage highlighted by Borges). According to Mill, if we did not have general names we could not state an attribute, express a simple comparison, or note any kind of regularity in nature. The "total idealism" of Tlön, which leads to the abolition of abstract ideas, brings to mind Ireneo Funes, who, following John Locke, set out to create a language in which each object—or, rather, each perception of each object—had its own name. Borges also reaffirms the importance of categorization in "The Analytic Language of John Wilkins" when he describes the bizarre and clearly useless classification of animals found in an apocryphal Chinese encyclopedia.

The abstraction process, the key to thought that reaches its pinnacle at the neurons in the hippocampus, is something we use every day. It is what a chess grandmaster needs to play a game and what I needed when I gave that talk in the auditorium at the University of Buenos Aires. Days before the talk I was unable (and unwilling) to remember every word or every sentence of what I was going to say. I tried to remember a general outline: I should talk about Borges first, then about Funes, and then about the brain and the Jennifer Aniston neurons. At a conference in Okazaki I saw several Japanese students memorize, word for word and in front of their advisors, the talks they were to deliver the next day. When the moment came, they repeated their talks like a recording but were unable to answer any questions, managing only to respond with generalities that they had also memorized. (Language limitations may have also played a role in this.) Verbal adhesion, memory without comprehension. But this is not intended as specific criticism of Japanese students, since many of us (not least in Argentina) tend sometimes to do the same because this is the way we are educated at school. We repeat and memorize for an exam but shortly thereafter we do not remember anything. We have already seen that we tend to forget details and remember concepts that set and solidify through repetition (reconsolidation) and association. This means that there is a serious general deficiency in the educational system. The problem has nothing to do with the subjects taught or the material imparted and is not the fault of either students or teachers; it has to do with the way in which the different topics are presented, studied, and tested. Students are bombarded with a deluge of facts and rarely have time to sit back and think, to extract the important ideas and link and contrast them to others;[14] in less than a couple of hours they go from Don Quixote to Pythagoras's theorem and from there to the French Revolution. On the other hand, if it is true that we eventually forget most of what we learn, what sense does it make to study? Why in the world did I spend hundreds of hours in the main auditorium of Pavilion I, if all I can remember is Paenza's explanation and a joke cracked by

14. As we saw in the previous chapter, the importance of creating associations in order to consolidate memories was noted by William James in the late nineteenth century. For example, it is much easier to remember that the battle of Chacabuco took place in 1817 once we know that Argentina declared independence the year before. Not only does this association help us remember a date, it also gives meaning by placing the date in context (i.e., right in the middle of the Latin American wars of independence).

an instructor whose name I forgot long ago? Luckily, those hours were not in vain because episodic memories (of events, of stuff that happens to us) give rise to semantic memories (of concepts). For example, I do not remember any of the lectures in which I was taught how to solve integrals but I do remember how to integrate. The episodes are long gone, but the concept remains.

We are reaching the end of our story. About a year ago I decided to follow Funes's trail to gain insights into Borges's brilliance in considering a problem that is at the forefront of today's neuroscience. I had the good fortune to end up working on very similar topics to those he described more than half a century ago, and this gave me the perfect excuse to immerse myself in his works and in those of the authors he read. Borges had the acuity to understand the importance of abstraction and oblivion, a theme that keeps coming back in Luria's description of Shereshevskii, the man who could not forget; in the experience with savants, who can remember infinitely many details but are incapable of simple reasoning; in the study of people with superlative memory; or in the abstract representations stored by neurons like Jennifer Aniston's in the hippocampus, which play a crucial role in memory.

In this twenty-first-century reality, in which we are constantly bombarded with information, where emails relentlessly alert us to all sorts of messages and we frantically follow the latest Facebook postings day to day, hour to hour, where news channels provide information 24 hours a day, seven days a week, and in mere seconds we switch from someone's tragedy to sports highlights and then to the weather forecast; in this cyberworld in which we must check and answer every text message as it arrives, where it is inconceivable to write a letter in longhand—partly because we barely know how to use a pen any more—where it is easier to meet someone in an online chatroom than in the real world; in this Funes kind of world, where the information overdose overwhelms us and we hardly have any time to think, where we barely perceive the passage of time and fate takes us through our lives and we hardly ever wonder why we do what we do; in the madness of our days, having these imaginary discussions with Borges and following his thoughts and his readings gave me the break I had long needed to step aside for a while and see things in perspective. This is the story that I wanted to share with you. As in the story mentioned in the introduction, the scientist—or perhaps the advisor to the ancient Persian king who finds in a tale told by a traveler the answer of a problem that has long kept him awake—takes a deep breath, makes himself comfortable in his chair, and feels the enormous satisfaction of having almost accomplished the task. I look for the perfect words to finish this book, I strive to close the circle that started with "Funes the Memorious" and then covered different aspects of what we know about memory and thought. The decision is easy: in this twenty-first century, amid the madness of an overwhelming reality, of something that goes far beyond our specific treatment of memory, the story of this *compadrito* from Fray Bentos is astonishingly contemporary. From my end, then, it only remains to step aside and quote the master:

Este, no lo olvidemos, era casi incapaz de ideas generales, platónicas. [. . .] Había aprendido sin esfuerzo el inglés, el francés, el portugués, el latín. Sospecho, sin embargo, que no era muy capaz de pensar. Pensar es olvidar diferencias, es generalizar, abstraer. En el abarrotado mundo de Funes no había sino detalles, casi inmediatos.

[Funes, let us not forget, was incapable of conceiving general, Platonic ideas. . . . He had effortlessly learned English, French, Portuguese, Latin. I suspect, however, that he was not very capable of thinking. To think is to forget differences, to generalize, to abstract. In Funes's crowded world there was nothing but almost immediate details.]

To Gerlind, Rodriguito, and Felipe
Leicester, November 2010

ACKNOWLEDGMENTS

The idea of this book originated from several conversations that I had with María Kodama. She suggested, in fact, that I write it. I wish to thank her for that initial spark, for her foreword, for her inclination to discuss these topics, and for repeatedly letting me study the books in Jorge Luis Borges's personal library.

It obviously takes a long time to write a book. I was lucky to be able to dedicate myself to it full-time (leaving many other tasks aside) thanks to a sabbatical at the University of Leicester and to a visiting professorship at the Department of Physics of the University of Buenos Aires in early 2010.

I would also like to thank Marcelo Panozzo, of Random House Mondadori, who strongly supported the original Spanish version of this book, and likewise Robert Prior and Susan Buckley at the MIT Press, who made this English translation possible. Borges, who was raised bilingual, joked that the Spanish version of *Don Quixote* was a bad translation from the English original. It may actually be the case that the English translation of this book surpasses the original. This shouldn't be attributed to my modest bilingual abilities, though, but to the terrific work and dedication of Juan Pablo Fernández.

Many of these pages were written at my in-laws' home in Germany and many others at my mother's house in San Isidro, among Bavarian beer and sausages, Argentine *asados* and red wine. I would like to thank them all for spoiling me—especially my *vieja*, who would wake up early to brew my *mate* while I pretended to be a writer; a luxury that one fully appreciates only after having been away for many years.

In this book I have used nontechnical language to touch upon several topics related to the brain's workings. At every stage there has been the danger, the temptation to oversimplify and renounce precision. To prevent this from happening (if I have managed to do so) I was lucky to count on comments and close readings by three brilliant colleagues: Gabriel Kreiman of Harvard University, Eduardo Mizraji of the Universidad de la República in Montevideo, and Horacio García of the Fundación para la Lucha contra las Enfermedades Neurológicas de la Infancia (FLENI), who also introduced me more than 20 years ago to the world of neuroscience and infected me with his curiosity about the brain and his passion to understand how it works. Several earlier drafts benefited greatly from the input of family and friends, who judged the book from a layperson's perspective. My most sincere thanks to Consuelo, Hugo, Tere, Isabel, Mariano, and Alejandro. Finally, I would not have been able to produce even a single page had I not had by my side a fabulous woman who kept supporting and encouraging me, and who always made sure I had all the time and the quiet that I needed to write. To her, and to the boys, thus, this book.

INDEX